# Carbonaceous Materials as Indicators of Metamorphism

*Edited by*

Russell R. Dutcher
Peter A. Hacquebard
James M. Schopf
Jack A. Simon

# Carbonaceous Materials as Indicators of Metamorphism

*Edited by*

Russell R. Dutcher
Department of Geology
Southern Illinois University at Carbondale
Carbondale, Illinois 62901

Peter A. Hacquebard
Geological Survey of Canada
Institute of Sedimentary and Petroleum Geology
Calgary, Alberta, Canada

James M. Schopf
U.S. Geological Survey
Ohio State University
Columbus, Ohio 43210

Jack A. Simon
Illinois Geological Survey
Natural Resources Building
Urbana, Illinois 61801

THE
GEOLOGICAL SOCIETY
OF AMERICA

SPECIAL PAPER 153

*Library of Congress Catalog Card Number 74-78627*
*I.S.B.N. 0-8137-2153-9*

Published by
THE GEOLOGICAL SOCIETY OF AMERICA, INC.
3300 Penrose Place
Boulder, Colorado 80301

*Printed in the United States of America*

*The printing of this volume has been made possible*
*through the bequest of Richard Alexander Fullerton Penrose, Jr.,*
*and the generous support of all contributors*
*to the publication program.*

# Contents

Foreword . . . . . . . . . . . . . . . . . . . . . . . . . . . . *Russell R. Dutcher* vii

Phytoclasts as indicators of thermal metamorphism, Franciscan assemblage and Great Valley sequence (upper Mesozoic), California . . . . . . . . . . *Neely H. Bostick* 1
- Abstract . . . . . . . . . . 1
- Introduction . . . . . . . . . . 2
- Phytoclast abundance and appearance . . . . . . . . . . 2
- Heat-induced phytoclast alteration . . . . . . . . . . 3
- Measurement of phytoclast alteration with the microscope . . . . . . . . . . 5
- Reference samples with known thermal history . . . . . . . . . . 7
- Great Valley sequence and Franciscan assemblage . . . . . . . . . . 8
- Organic rank of Great Valley sequence samples . . . . . . . . . . 11
- Organic rank of some Franciscan rocks . . . . . . . . . . 12
- Relation between organic rank of Great Valley sequence and Franciscan rocks 13
- Acknowledgments . . . . . . . . . . 15
- References cited . . . . . . . . . . 16

Microscopic examination of kerogen (dispersed organic matter) in petroleum exploration . . . . . . . . . . *Jack D. Burgess* 19
- Abstract . . . . . . . . . . 19
- Introduction . . . . . . . . . . 20
- Source rocks . . . . . . . . . . 21
- Kerogen usage . . . . . . . . . . 21
- Thermal changes in kerogen . . . . . . . . . . 22
- Types of organic matter . . . . . . . . . . 24
- Examples of kerogen study in oil exploration . . . . . . . . . . 26
- Summary . . . . . . . . . . 27
- Acknowledgments . . . . . . . . . . 29
- References cited . . . . . . . . . . 29

Interpretation of vitrinite reflectance measurements in sedimentary rocks and determination of burial history using vitrinite reflectance and authigenic minerals . . . . . . . . . . *John R. Castaño and Dennis M. Sparks* 31
- Abstract . . . . . . . . . . 31
- Introduction . . . . . . . . . . 32
- Interpretation of vitrinite reflectance data in non-coal-bearing sequences . . . . 32
- Factors governing coalification . . . . . . . . . . 35
- Determination of thermal history using authigenic minerals and vitrinite reflectance 36

Factors governing the formation of laumontite . . . . . . . . . . . . . . . . . . . 49
Conclusions . . . . . . . . . . . . . . . . . . . . . . . . . . . . . . . . . . . . . . 49
Acknowledgments . . . . . . . . . . . . . . . . . . . . . . . . . . . . . . . . . . . 50
References cited . . . . . . . . . . . . . . . . . . . . . . . . . . . . . . . . . . . 51

Coalification patterns of Pennsylvanian coal basins of the eastern United States
. . . . . . . . . . . . . . . . . . . . . . . . . . . . . . . . *Heinz H. Damberger* 53
Abstract . . . . . . . . . . . . . . . . . . . . . . . . . . . . . . . . . . . . . . . 53
Introduction . . . . . . . . . . . . . . . . . . . . . . . . . . . . . . . . . . . . . 54
Regional coalification patterns of Pennsylvanian coal basins of the United States 56
Acknowledgments . . . . . . . . . . . . . . . . . . . . . . . . . . . . . . . . . . . 70
Appendix 1. Collection and manipulation of data for coalification maps . . . . 71
References cited . . . . . . . . . . . . . . . . . . . . . . . . . . . . . . . . . . . 72

Rank studies of coals in the Rocky Mountains and inner foothills belt, Canada
. . . . . . . . . . . . . . . . . . . . . . . . . . . *P. A. Hacquebard and J. R. Donaldson* 75
Abstract . . . . . . . . . . . . . . . . . . . . . . . . . . . . . . . . . . . . . . . 75
Introduction . . . . . . . . . . . . . . . . . . . . . . . . . . . . . . . . . . . . . 76
Rank parameter employed . . . . . . . . . . . . . . . . . . . . . . . . . . . . . . 76
Regional rank changes and effect of tectonism . . . . . . . . . . . . . . . . . . 79
Rank-depth relations . . . . . . . . . . . . . . . . . . . . . . . . . . . . . . . . . 83
Seam correlations based on rank obtained from vitrinite reflectance measurements 90
Acknowledgments . . . . . . . . . . . . . . . . . . . . . . . . . . . . . . . . . . . 92
References cited . . . . . . . . . . . . . . . . . . . . . . . . . . . . . . . . . . . 93

Vitrinite reflectance as an indicator of coal metamorphism for cokemaking
. . . . . . . . . . . . . . . . . . . . . . . . . . . . *R. R. Thompson and L. G. Benedict* 95
Abstract . . . . . . . . . . . . . . . . . . . . . . . . . . . . . . . . . . . . . . . 95
Introduction . . . . . . . . . . . . . . . . . . . . . . . . . . . . . . . . . . . . . 96
Measuring reflectance in relation to the pseudovitrinite-vitrinite subdivision . . 97
Discontinuities in the coal metamorphic series . . . . . . . . . . . . . . . . . . 102
Conclusion . . . . . . . . . . . . . . . . . . . . . . . . . . . . . . . . . . . . . . 107
References cited . . . . . . . . . . . . . . . . . . . . . . . . . . . . . . . . . . . 108

# Foreword

At the 1969 annual meeting of the Coal Geology Division of The Geological Society of America, initial plans were made to hold a symposium on the metamorphosis of coaly material. This symposium, entitled "Carbonaceous Materials as Indicators of Metamorphism," was held at the Milwaukee meetings in 1970. The papers contained in this volume were presented at that time.

In the late 1960s, many scientists, both in and out of industry, became interested in the increases in rank exhibited by organic matter, whether finely dispersed in shales and sandstones or present in coal seams. It was suggested that this idea was possibly of great significance in the exploration for oil and gas. The immense importance of coal rank in the manufacture of metallurgical coke has long been recognized, and the steel industry has made great strides not only in understanding but also in predicting coal behavior. Techniques were refined, and some techniques that were developed in this search for better coke are now being applied to other organic matter trapped in sedimentary rocks in an attempt to predict or delimit "gas"-"no gas" zones or "wet and dry" areas in petroleum exploration.

Many scientists thought this was all new—others recognized that we, like "continental drifters," were really returning to ideas that had been presented many years ago. Over one hundred years ago, Henry Darwin Rogers noted the relation between coal rank and petroleum distribution. David White further defined the carbon ratio theory in 1915 when he stated:

> Wherever the regional alteration of the carbonaceous residues passes the point marked by 65 percent or perhaps 70 percent of fixed carbon in the (pure) coals, the light distillates appear, in general, to be gases at rock temperature.

Our present efforts are in a true sense merely an extension of this idea. Admittedly, we have used some new techniques and manipulated data in different fashions, but we are here, as in most geological work, building upon the work of others, for which they should receive credit.

The authors of the papers included in this volume have been a great help throughout the lengthy editing process. The wide variety of interests represented gives some indication of the breadth of coal petrologic and petrographic studies today.

My sincere thanks and appreciation, and those of the entire Coal Geology Division, are extended to Peter A. Hacquebard of the Geological Survey of Canada, James

M. Schopf of the U.S. Geological Survey, and Jack A. Simon of the Illinois Geological Survey. They have all devoted much time to the planning, execution, and editing associated with the symposium and this resultant volume. We hope that the papers presented here will be of interest and possibly of some value to those interested in the effect of rank change upon coal behavior and of the effect of the processes that produced these changes upon other enclosing sediments.

RUSSELL R. DUTCHER

*Professor and Chairman*
*Department of Geology*
*Southern Illinois University at Carbondale*
*Carbondale, Illinois 62901*

Geological Society of America
Special Paper 153


# Phytoclasts as Indicators of Thermal Metamorphism, Franciscan Assemblage and Great Valley Sequence (Upper Mesozoic), California

NEELY H. BOSTICK

*Coal Section*
*Illinois State Geological Survey*
*Urbana, Illinois 61801*

## ABSTRACT

Grains of clastic organic matter similar to coal occur scattered among the mineral grains in most shale and sandstone; they are called phytoclasts because many of them have relict plant structures. The phytoclasts that have been least altered in a given sample can be used to determine the degree of metamorphism of the host rock.

As the rank of phytoclasts increases after burial—mainly in response to temperature increase—rank can be used to indicate past rock temperature in the range of about 80° to 200°C for exposure times longer than those associated with contact metamorphism.

The proportions of different phytoclasts vary greatly in different samples, making bulk chemical analysis unreliable for determining rank. Optical properties of phytoclasts can be used as parameters of rank, and measurement of reflectance of polished grain mounts under the microscope is the technique applicable to the greatest variety of samples and best suited to selection of the least altered grains in a sample.

Reflectance of phytoclasts from drill-hole cores, the duration of phytoclast burial in the host rocks, and the present-day temperatures of the rocks are used to construct a rank-time-temperature model of organic metamorphism. The model is used for the Great Valley sequence and Franciscan assemblage, thick upper Mesozoic strata in California. Reflectance of phytoclasts from rocks of the Great Valley sequence that are now exposed increases with interpreted former burial depth, but the deduced past temperature gradient was unusually low, about 5°C per km (0.17°C per 100 ft). This low gradient value also appears to apply to the Franciscan assemblage where it is in thrust contact underneath the Great Valley strata.

The past temperature, thermal gradient, and burial depth of Franciscan rocks, interpreted from the gradient of phytoclast alteration, corroborate studies of mineral paragenesis in these rocks, which are in the zeolite and blueschist metamorphic facies. The former low thermal gradient and the structural position of these strata support the concept of crustal tectonics in which the Franciscan assemblage moved under coeval Great Valley sediments, and both subsequently rose rapidly and were dissected by erosion.

## INTRODUCTION

Most sedimentary rocks contain dispersed organic particles, carbon-oxygen-hydrogen-nitrogen compounds like those composing coal, that are about the same size as the clastic mineral grains of the rock. In ordinary shales and sandstones, most of these particles appear to have been derived from plants and to have been transported and deposited through normal sedimentary processes; hence, they have been named phytoclasts (Bostick, 1971).

Phytoclast metamorphism in sedimentary rocks is mainly a response to increased temperature. Described in this paper are (1) changes that phytoclasts exhibit during natural and laboratory thermal metamorphism, (2) the reference samples used for determination of absolute past temperature, and (3) application of this technique of paleothermometry to Mesozoic rocks in California that once were deeply buried.

In the laboratory, the phytoclasts are concentrated by crushing rock samples, macerating them with HCl and HF, and eliminating remaining minerals by floating the organic particles on a liquid with 2.1 specific gravity.

## PHYTOCLAST ABUNDANCE AND APPEARANCE

Organic particles are a ubiquitous constituent of sedimentary rocks, although they usually make up less than one percent of the rock mass and are generally ignored by petrologists. The Cretaceous shales and siltstones of California that were studied contain 0.1 to 0.4 weight percent organic particles recovered after thorough maceration.

Publications on the organic constituents of sedimentary rocks usually give quantitative data as percentage of *total* organic carbon from combustion analysis, and conversion from percentage of organic carbon to percentage of organic matter is only approximate. At least 90 to 95 percent of the organic matter is phytoclasts rather than compounds soluble in organic solvents (Schrayer and Zarella, 1966; Stevenson and Dickerson, 1969); so combustion analysis data by various authors show, roughly, the percentage of phytoclasts to be expected in rocks of different lithology and different geologic age. Figure 1 shows the amount of organic matter in three broad classes of sedimentary rocks (Ronov, 1958). The present study suggests modification of this picture, for red and green claystones or sandstones contain much less organic matter than is shown as average for rocks of that grain size, and the amount of organic matter in carbonates is extremely variable. Figure 2 shows the amount of organic matter in sedimentary rocks of different ages in the United States and the Russian Platform area.

Phytoclasts viewed under the microscope by transmitted light on a strew mount as is usually used in palynology are commonly too thick and opaque to be identified (Fig. 3a); however, low-rank samples reveal identifiable spores and pollen, along with other particles thin enough for internal structure to be seen (Fig. 3b). Phytoclasts can also be embedded in plastic and studied as polished sections with a reflected-light microscope (Fig. 4). With this technique, internal structure shows even in opaque high-rank material, and the phytoclasts are seen to consist of plant spores and pollen, leaf cuticles, and fragments that have bordered pits, ribs, fibers, or cellular structure. Apparently unstructured, compact particles also are revealed that probably are equivalent to the collinite described by coal petrologists or to insoluble humate reported by geochemists. In addition, the insoluble organic residues of some organic-rich rocks are shown to contain irregular or flocculent masses, some of which may be of algal origin.

## HEAT-INDUCED PHYTOCLAST ALTERATION

The visible and chemical changes phytoclasts exhibit when they have been heated are like those described for coal. Thin grains under transmitted light are seen

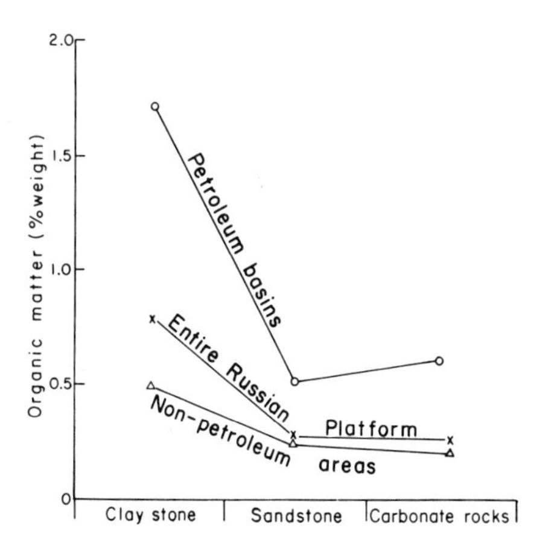


**Figure 1. Average amount of organic matter in sedimentary rocks of different lithology from petroliferous and nonpetroliferous areas of the Russian Platform, based on Ronov (1958). Values converted from organic carbon percentage by using an organic factor of 1.22, determined by Forsman and Hunt (1958).**

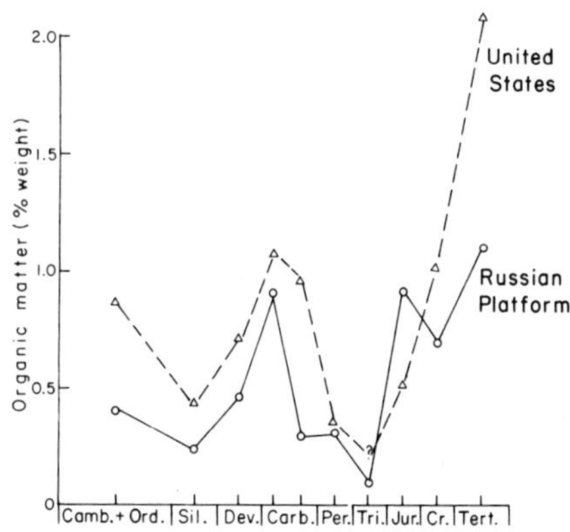


**Figure 2. Variations in the average amount of organic matter in sedimentary rocks from different geologic systems. Adapted from data on organic carbon content of 7,050 samples by Trask and Patnode (1942) (mainly from petroleum basins) and 25,700 samples by Ronov (1958) by using the organic factor 1.22.**

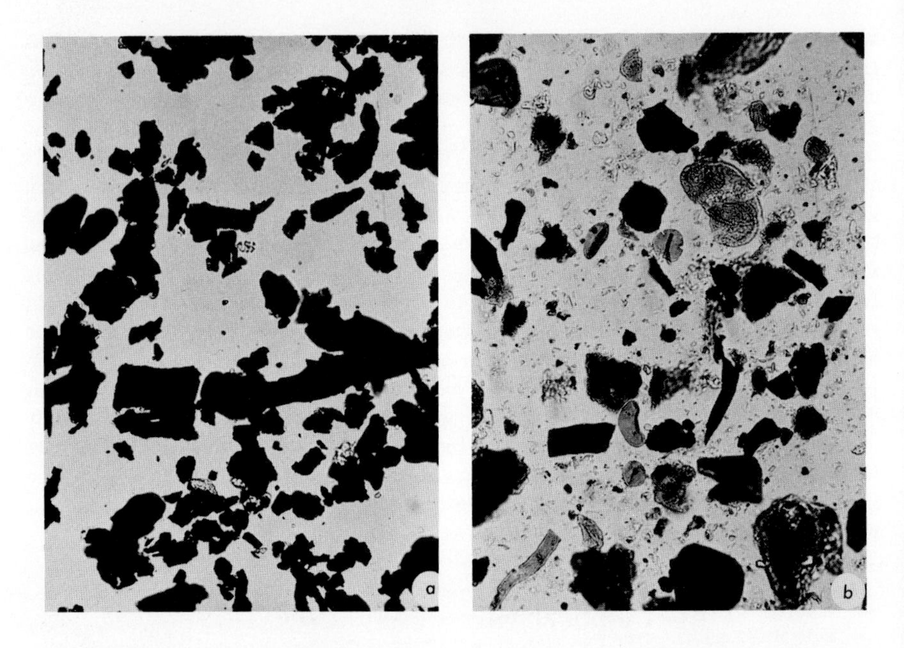

**Figure 3. Phytoclast assemblages in strew mounts of whole grains, viewed by transmitted light microscope. The field greatest dimension is 0.53 mm. (a) Franciscan assemblage, upper unit, Diablo Range, from a zone where graywackes contain jadeitic pyroxene. See Figure 4a. (b) Eocene siltstone, La Jolla, California; 30 ft from the contact with a 10-ft-thick basalt dike.**

to change hue from pale yellow through orange to gray-brown, and the grains look tattered or porous. Under reflected light, polished grain surfaces look brighter the more they have been exposed to heat, and some larger compact grains develop micropores.

Ultimate chemical analyses of phytoclast concentrates from rocks that have not been greatly heated are similar to those of lignite or subbituminous coal, while analyses from rocks that have been deeply buried or near igneous contacts are similar to those for higher rank coal (Bostick, 1970). In response to increased heating, relative carbon, by weight, increases from about 65 to 90 percent, while oxygen, hydrogen, and nitrogen all decrease, typically from 27 to 7 percent, 6 to 2.5 percent, and 2 to 0.5 percent, respectively.

These figures are mentioned here because such analyses are a means of determining rank that is familiar to coal scientists, and phytoclasts are basically coal particles. However, there are obstacles to the use of chemical analyses for expressing phytoclast rank. The heterogeneous phytoclast concentrates from different rocks may consist partly of material redeposited from much older strata and also may contain a greatly variable content of such hydrogen-rich grains as spores and plant cuticles or carbon-rich fusinitic grains. Also, combustion chemical analysis is sometimes difficult, because mineral constituents cannot be reduced to a level where they do not affect analysis, and because low-rank phytoclast concentrates are strongly hygroscopic.

## MEASUREMENT OF PHYTOCLAST ALTERATION WITH THE MICROSCOPE

Studies with the microscope show the optical properties of single particles and thus allow particular types to be selected from the phytoclast assemblage of a sample. To be most useful as indicators of thermal history of the host rock, the material selected should have optical properties that can be measured quantitatively and should be common to all samples, easily recognized, not redeposited from older rocks, and not strongly altered biochemically.

Plant spores are useful indicators, especially for studies with transmitted light, but they are sparse in many samples, or even absent, and may be unidentifiable in metamorphosed rocks. Vitrinitic material is generally too thick and opaque for its properties to be measured in transmitted light, but it is very useful in reflected light. It occurs in virtually all samples and is generally compact and easily polished. Its reflectance increases regularly with increased thermal alteration, and it has been studied extensively by coal petrologists (de Vries and others, 1968). The problem of selecting vitrinitic particles whose rank represents the thermal history of the host rock is illustrated schematically in Figure 5. A sediment can contain organic particles derived directly from plants, particles somewhat altered biochemically during transport from plant to final sediment burial, and particles from older sedimentary rocks that have been considerably metamorphosed. To measure

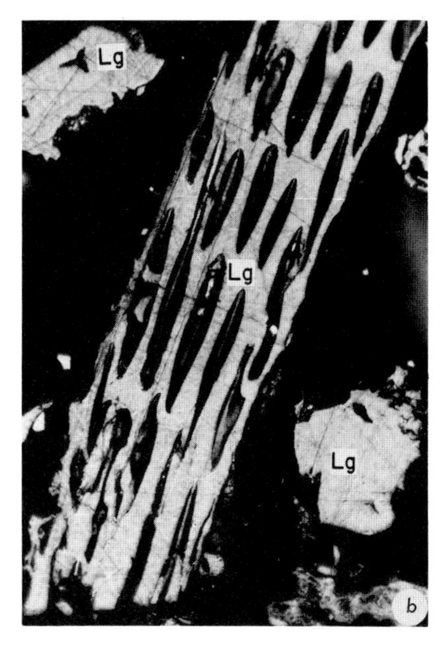


**Figure 4. Phytoclast assemblages in polished section, viewed by reflected light microscope. Lg marks grains of low-gray vitrinite. (a) Same sample as in Figure 3a. Reflectance of low-gray vitrinite grains is 1.3 percent in oil. The field greatest dimension is 0.36 mm. (b) Colorado shale, Upper Cretaceous; Cascade County, Montana; 2 ft from a dike about 25 ft thick. Reflectance of low-gray vitrinite grains is about 1.8 percent in oil. The field greatest dimension is 0.15 mm.**

alteration reflecting the history of the present host rock, those particles that had the lowest degree of alteration when incorporated in the sediment being studied must be selected.

Spores, leaf cuticles, and resin are the phytoclasts with the lowest reflectance, and they can also be recognized by their distinctive forms. The remaining particles, which are apparently equivalent to vitrinite, semifusinite, and fusinite in coal, have diverse reflectances in a sample. When polished grain mounts are used, a class of particles, called here "low-gray" (Lg in Fig. 4), can be separated under the microscope, and the reflectance can be measured with a photomultiplier photometer. As this low-gray is apparently the vitrinite which had lowest rank

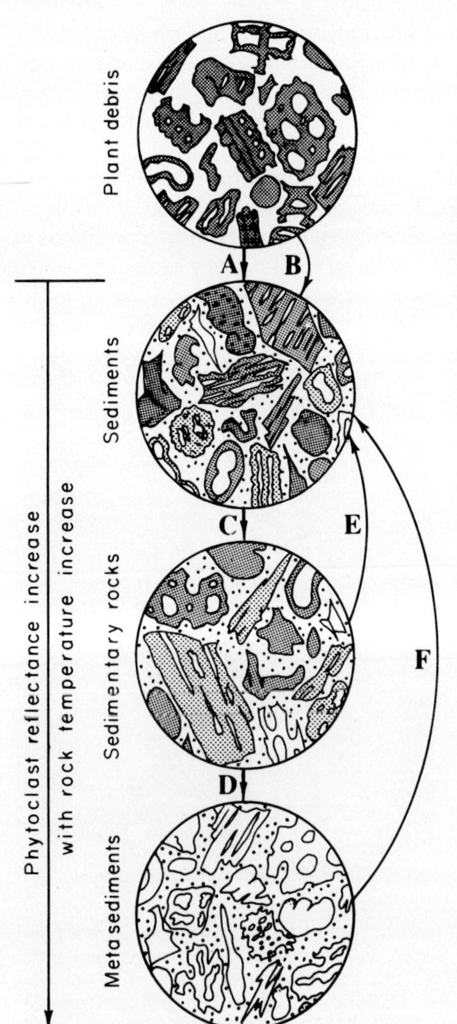


**Figure 5. Schematic representation of sedimentary rocks as seen by reflected light, showing the relation between phytoclast alteration and host rock metamorphism. Phytoclasts are actually the same size as mineral grains but are exaggerated in the figure. (A) Plant material is deposited with mineral grains. (B) Some plant material is biochemically altered during delayed transport from plant to sediment. (C) Sediment is buried and indurated to form rock. (D) Sediment is further buried and phytoclasts acquire high reflectance. Phytoclasts (except fusinite) respond to increased burial temperature, as can be seen by increased reflectance (lighter tone in figure). (E), (F) Some phytoclasts are eroded and redeposited in younger sediments; they already have relatively high rank when redeposited. The best measure of temperature increase from sediment to rock is the reflectance increase of the least altered phytoclasts in the given sample, which have not been redeposited or highly altered before burial.**

when the sediment formed, its present rank must have developed within the host rock. In many phytoclast assemblages there is abundant, uniform low-gray material; in others it is scarce, and the microscopist must work laboriously to locate low-gray particles for reflectance measurement from a heterogeneous assemblage or else measure many particles at random and select the lowest values of the total assemblage (excluding spores and resin) as a measure of rank developed in the rock.

## REFERENCE SAMPLES WITH KNOWN THERMAL HISTORY

Analysis of phytoclasts found in suites of shale and siltstone from igneous dike contact zones and from deep wells showed that their rank increases regularly with increased rock temperature. Low-gray reflectance analyzed with an electronic photometer provides a quantitative measure of the alteration state. Two approaches are available to relate the alteration to actual rock temperatures: laboratory bomb experiments with good control of conditions, and field studies, in which measured rock temperatures are less certain.

The literature on coking contains many studies of coal thermal alteration in the laboratory, and curves for five of these for which reflectance data are available are plotted on Figure 6. The samples used in these studies were coal samples, nearly mineral free, that were heated dry in air or nitrogen for 20 hrs at most, all but one at atmospheric pressure. Clearly, these conditions are far from the natural conditions of buried phytoclast samples. In the present study, natural conditions were more nearly matched with a series of bombs containing argillaceous lignite and water vapor sealed in gold. They were heated under the pressure usually found in sediments buried at the temperature of the run. Conditions were kept constant for 30 days.

The results of the bomb experiments also are plotted in Figure 6. They are surprisingly parallel to the published results for true coals altered under the unnatural laboratory conditions.

The curve of phytoclast reflectance and temperature on Figure 6 serves as a thermometer to relate alteration to past rock temperature. This thermometer was tested with phytoclasts from samples of shale collected adjacent to volcanic dikes, most of them from a single layer of rock at measured distances normal to the dike. The phytoclast reflectance values from these contact samples were converted to temperature values by using the bomb-sample curve of Figure 6, and the results are plotted in Figure 7. Temperature increases regularly toward the contact, and, furthermore, the curves for the suites of samples have similar shapes. The curves from phytoclast analysis can be compared with curves of country rock temperatures adjacent to a dike based on heat-flow theory. A curve that shows theoretical maximum rock temperature in rocks similar to the samples used for a given distance away from an igneous contact is also plotted on Figure 7 (heavy broken line). The temperature curves derived from heat theory and from phytoclast analysis are strikingly similar.

The heating time for rocks adjacent to dikes a few meters thick is on the order of a month or two (Lovering, 1935), so the element of time for the dike situation corresponds fairly well with that for the hydrothermal bomb experiments. To study thermal alteration resulting from longer heating, the rank of phytoclasts was measured from drill-hole samples for which present rock temperature had been measured. As shown in Figure 8, the reflectance increases more for a given temperature increase for drill-hole samples than for the one-month hydrothermal bomb samples. For suite O (Carboniferous rocks), the present rock temperature may not have been the maximum to which the strata were subjected, but the data form a consistent pattern. The well suites on which curves W and E were based contain rocks buried since the Jurassic and Cretaceous periods, and a smooth "well-standard" line has been plotted in Figure 8 with the same rank-temperature progression. This line was used as an approximate temperature predictor for burial metamorphism of Cretaceous and Jurassic strata in California, as is described below.

## GREAT VALLEY SEQUENCE AND FRANCISCAN ASSEMBLAGE

Upper Jurassic and Cretaceous marine sandstones and mudstones are exposed in much of central California west of the Great Valley (Fig. 9). They have been studied in detail in the Lower Lake-Cache Creek area northwest of Sacramento

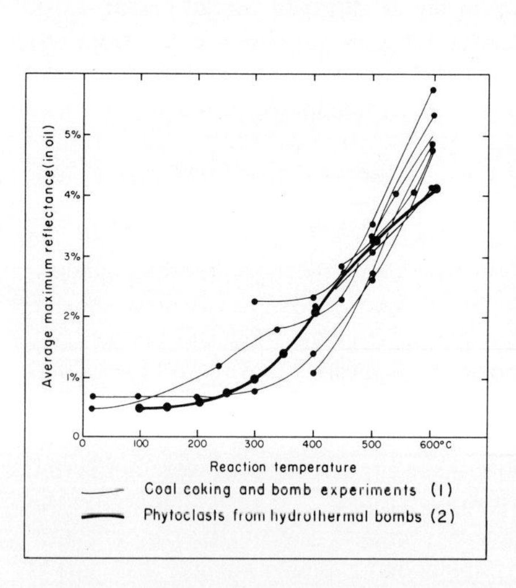


**Figure 6. Vitrinite reflectance and reaction temperature of (1) coal heated an hour or two in coking experiments (Hryckowian and others, 1967; Chandra and Bond, 1956; Wang, 1963; Brown and Taylor, 1961; Ghosh, 1967; Kisch, 1966), and (2) lignitic shale saturated with water vapor and heated one month under pressure.**

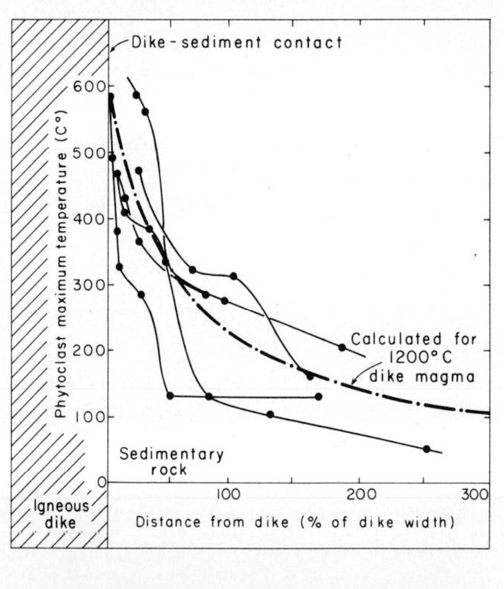


**Figure 7. Maximum past temperature in five igneous dike contact aureoles in Cretaceous and Eocene shales and siltstones, based on phytoclast reflectance converted to temperature by using the reflectance-temperature curve for hydrothermal bomb samples in Figure 6. The dikes are 12 to 30 ft thick. The heavy broken line, from heat-flow studies by Lovering (1935) and Jaeger (1959), shows the theoretical maximum temperature that the country rock reached at any time after intrusion, in respect to the dike contact, assuming abrupt intrusion rather than continuous flow.**

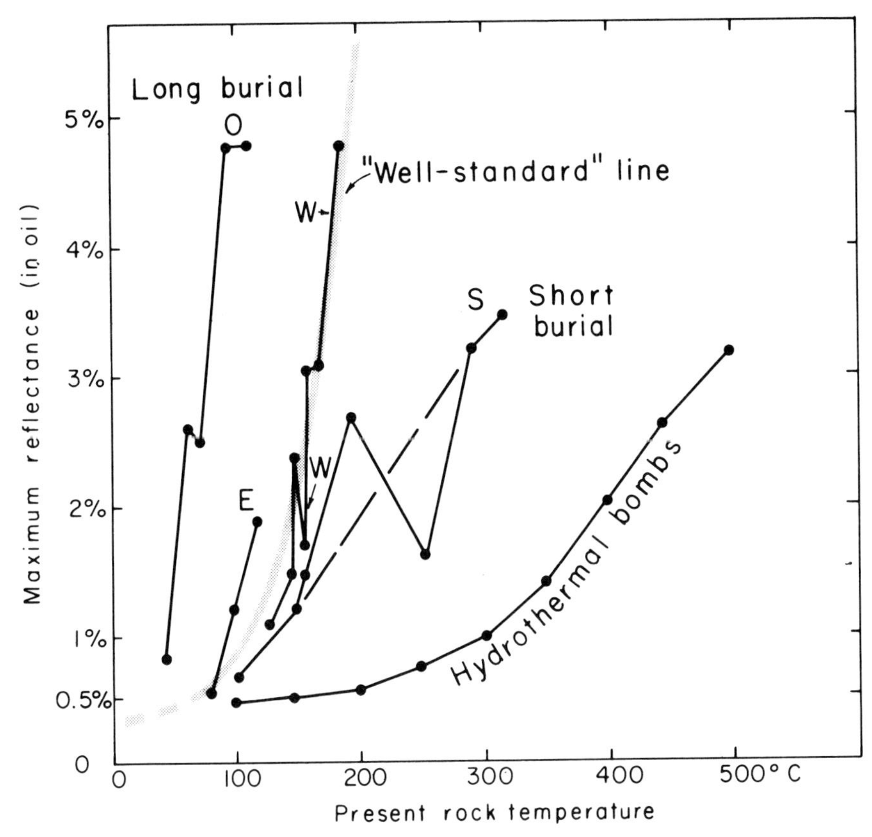


**Figure 8. Phytoclast reflectance and present rock temperature measured for samples from deep wells. Suite O contains Carboniferous rocks from the Arkoma Basin, Oklahoma. W and E are Jurassic and Lower Cretaceous rocks from Texas, and S is a well in the Salton Geothermal Field, California, Pleistocene and Pliocene. Also plotted are data from phytoclasts heated one month in hydrothermal bombs. The "well-standard" line is used to estimate former temperature from reflectance of samples buried roughly 30 to 80 m.y.**

(Dickinson and others, 1969; Swe and Dickinson, 1970), where their total thickness is about 35,000 ft (11 km). These strata are referred to informally in this paper as the Great Valley sequence.

A heterogeneous Upper Jurassic and Cretaceous assemblage of shales and graywackes containing some chert, limestone lenses, and volcanic rocks occupies generally the same terrain in central California as the Great Valley sequence (Figs. 9 and 10). Sedimentary rocks in this assemblage are generally much more compacted, disrupted, and metamorphosed than those of the Great Valley sequence, with many sites of blueschist facies metamorphism. Bailey and others (1964, p. 11) suggested that these metamorphosed rocks be called the Franciscan assemblage. At the known exposures of contact between Franciscan and Great Valley strata in this area, the Franciscan is in thrust contact beneath the Great Valley sequence or is separated from it by a layer of serpentine.

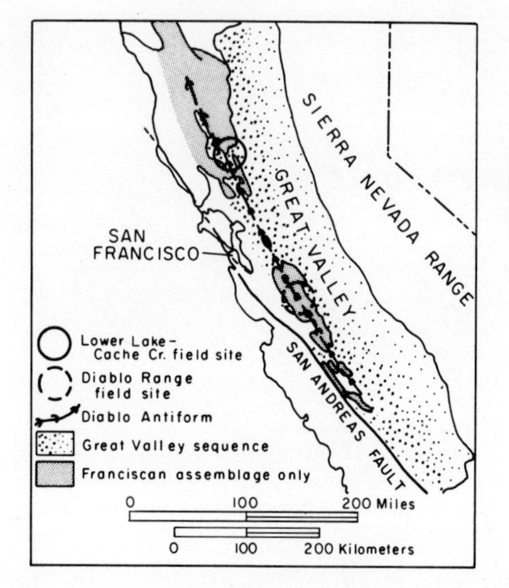


**Figure 9. Location map of central California showing the field study areas. The trend of the Diablo Antiform is from Bailey and others (1964, p. 150); it is not a continuous structure.**

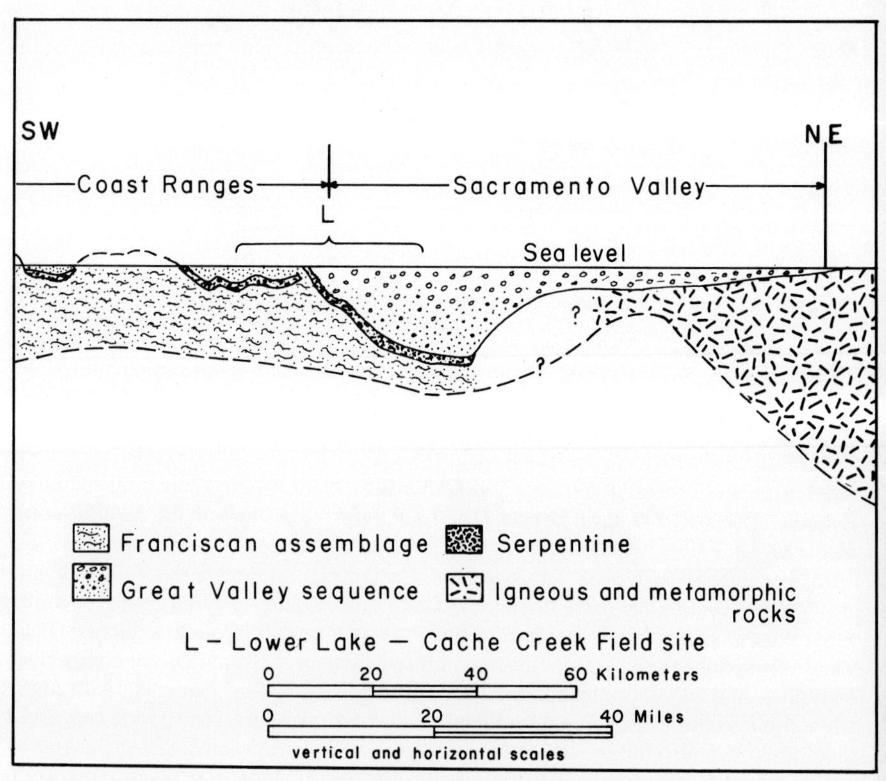


**Figure 10. Idealized vertical section across the Sacramento Valley and part of the coast ranges near the Lower Lake–Cache Creek Field area (after Bailey and others, 1964, p. 164).**

## ORGANIC RANK OF GREAT VALLEY SEQUENCE SAMPLES

Five samples of Great Valley rocks from the Lower Lake-Cache Creek area, for which the stratigraphic control was good, were analyzed by phytoclast reflectance measurement. The expected increase of rank with burial depth was confirmed, but reflectance was unexpectedly and consistently low for all the samples (Table 1). The estimated former burial depth of the deepest samples was 36,000 ft (11 km), and the temperature of rocks at this depth would be 235°C with a low thermal gradient of 0.6°C per 100 ft (18°C per km), which is about half the average in sedimentary basins today. Yet from the measured reflectance, converted by the control curve from well samples (Fig. 8), the temperature of the deepest samples is found to be 82°C, and could be perhaps 100°C at most.

In the two right-hand columns of Table 1 are plotted rock temperatures based on sample depths and two assumed thermal gradients. Temperatures for the two shallow samples are uncertain because the rank-temperature control curve is poorly defined for low rank. With the exception of these two samples, an assumed low gradient of 0.6°C per 100 ft (18°C per km) gives rock temperatures still much higher than those found from phytoclast reflectance analysis. Only with an extremely low gradient of about 0.17°C per 100 ft (5.6°C per km) do calculated temperatures and temperatures from measured organic metamorphism agree; this gradient is about one-sixth of normal gradients and only one-third of the lowest gradients now measured at the Mississippi Delta or in southern Florida (Moses, 1961; Griffin and others, 1969).

Besides the paleotectonic model with low thermal gradient, which is discussed at the end of this paper, several explanations for the unusually low reflectance values were considered. As shown in Figure 8, phytoclast reflectance increases with burial time. The "well-standard" line is plotted to accord approximately with data for well samples that had minimum burial time of 40 m.y. at the present depth and 65 m.y. at about two-thirds the present depth. Could very brief burial account for the low reflectance of the Lower Lake-Cache Creek samples? Study of the geologic setting suggests at least a 20 m.y. burial time for the youngest samples (Swe and Dickinson, 1970, p. 183). As shown in Figure 8, samples with 0.6 percent reflectance, the highest value found in the Cache Creek study, could not have persisted at a temperature over 100°C, even for as short a period as that represented in well S, which intersected strata from Pleistocene to present age.

Another explanation for the very low reflectance values is based on the argument that the depth of former burial given for these samples is much too great. The lowest (Jurassic) samples (see L on Fig. 10) were collected some distance west of the highest (Upper Cretaceous) samples, and deposition might have shifted eastward with time so that Jurassic strata were not covered by the full sequence of Cretaceous strata. However, this is not believed to be the case; several authors have presented field evidence that the 35,000-ft stratigraphic thickness of the section is nearly equal to former burial depth plus 5,000-ft burial under Tertiary strata (see references in Dickinson and others, 1969).

In summary, the Great Valley sequence samples studied contain phytoclasts that have increased metamorphic rank with increased depth in the section, but the temperature values from reflectance measurements are unusually low and indicate

Table 1. Data from Reflectance of Phytoclasts in the Great Valley Sequence Samples from the Lower Lake–Cache Creek Section

| | Age and sample no. | Estimated former depth (ft) | Modal maximum reflectance of low-gray phytoclasts (%) | Temp. from "well line" (°C) | Temp. based on depth and assumed gradient (°C) 0.6°C/100 ft (18°C/km) | 0.17°C/100 ft (5.6°C/km) |
|---|---|---|---|---|---|---|
| Upper Cretaceous | Campanian 372 | 2,000 | 0.40 | 30 | 32 | 23 |
| Upper Cretaceous | Campanian-Maestrictian 344 | 6,500 | 0.45 | 45 | 56 | 31 |
| Lower Cretaceous | Hauterivian 371 | 32,000 | 0.50 | 70 | 212 | 74 |
| Upper Jurassic | Tithonian 373 | 36,000 | 0.60 | 82 | 235 | 81 |
| Upper Jurassic | Tithonian 346 | 36,000 | 0.60 | 82 | 235 | 81 |

Note: Age and burial depth are from Swe and Dickinson (1970).

that the thermal gradient in these strata did not rise above 5.6°C per km, one-sixth normal, from the time of deposition until the strata were tilted and eroded in the late Eocene.

## ORGANIC RANK OF SOME FRANCISCAN ROCKS

The Central Diablo Range, about 100 mi southeast of San Francisco (Fig. 9) is a Franciscan locality where the gross structure appears to be an anticline and the strata are less broken and distorted than usual. This Franciscan structure has been termed the Diablo Antiform and the strata have been subdivided into an upper and lower unit (Soliman, 1965, p. 24). Phytoclast reflectance analysis of eight samples taken along the Mount Hamilton road across the antiform supports the ideas of upper and lower units and anticlinal structure. As four samples in the upper unit have a phytoclast oil-immersion reflectance Ro of 1.1 to 1.3 percent, and three in the lower unit have Ro 1.7 to 2.1 percent, it appears that the lower unit has been hotter, presumably deeper. The gap in reflectance range of the two units suggests that they are not in simple depositional contact, although there is considerable scatter of values for each unit.

Six additional samples from the Garzas Creek area, about 20 mi southeast of Mount Hamilton on the east side of the Diablo Range, yielded reflectance values of the same magnitudes. Two relatively "hot" samples (Ro of 1.7 to 1.8) are from chaotically disturbed strata, three relatively "cool" samples (Ro of 1.2 to 1.3) are from coherent metagraywacke, and the "coolest" sample (Ro of 1.1) is from jadeitic pyroxene-bearing metagraywacke sampled 50 m from the eastern boundary of the Franciscan outcrop. The equivalent temperatures from the "well-standard" curve (Fig. 8) range from 140° to 150°C for the "hot" group of samples and from 115° to 130°C for the "cool" group (Fig. 11). As burial depth of these samples is probably about 50,000 ft (Bailey and others, 1964, p. 21), these temperatures are much too low unless the geothermal gradient was very small (or burial time was shorter than one million years). Also indicative of a very small gradient is the short range of phytoclast reflectance values found within these Franciscan rocks despite their considerable thickness and structural complexity.

## RELATION BETWEEN ORGANIC RANK OF GREAT VALLEY SEQUENCE AND FRANCISCAN ROCKS

The sample sites of Franciscan and Great Valley rocks described above are about 120 mi apart, but Great Valley strata overlie Franciscan rocks in the same structural setting at both sites along the trend of the Diablo Antiform (Fig. 9) (Bailey and others, 1964, p. 150). Despite the considerable distance along strike between these localities, a plot of the temperature-depth data for the two suites of rocks on a single temperature-pressure diagram tentatively relates their burial and temperature histories (Fig. 11). Burial depth and maximum past temperature for the Great Valley samples, both determined in this study, are plotted directly on the diagram. If the same geothermal gradient is assumed for both suites, the Franciscan rocks can be plotted on the gradient line from their former temperature alone, from phytoclast reflectance, to show their likely former burial depth. The resulting depth of burial for Diablo Range Franciscan rocks is 18 to 19 km (about 60,000 ft) for the "cool" (flank) samples and 21 to 23 km (about 72,000 ft) for the "hot" (core) samples.

These new temperature data for Diablo Range Franciscan rocks and the derived former maximum depth of burial from extension of the "fossil" geothermal gradient in Great Valley strata invite comparison with information about paragenesis of metamorphic minerals in the rock. Most graywacke interbedded with the mainly shale samples analyzed by phytoclast reflectance contains lawsonite, and the two samples from near the margin faults of the Diablo Range contain jadeitic pyroxene (M. C. Blake, Jr., and D. S. Cowan, 1970, oral commun. ).

The presence of lawsonite in Franciscan samples is in accord with P-T data shown on Figure 11 at about a 10-km depth. The line marking the P-T boundary between rocks with and without jadeitic pyroxene conforms with the circumstance that the Franciscan samples contain albite but are free of jadeitic pyroxene. The two exceptions, containing jadeitic pyroxene, are "cold" samples collected near the margin of the range; with positions indicated by arrows on Figure 11, these two samples could have reached conditions for growth of jadeitic pyroxene if the pressure had been increased one kilobar locally above regional pressure.

The experimental aragonite-calcite line lies just above the position of the Diablo Range Franciscan samples. Some of the Franciscan rocks contain aragonite, and

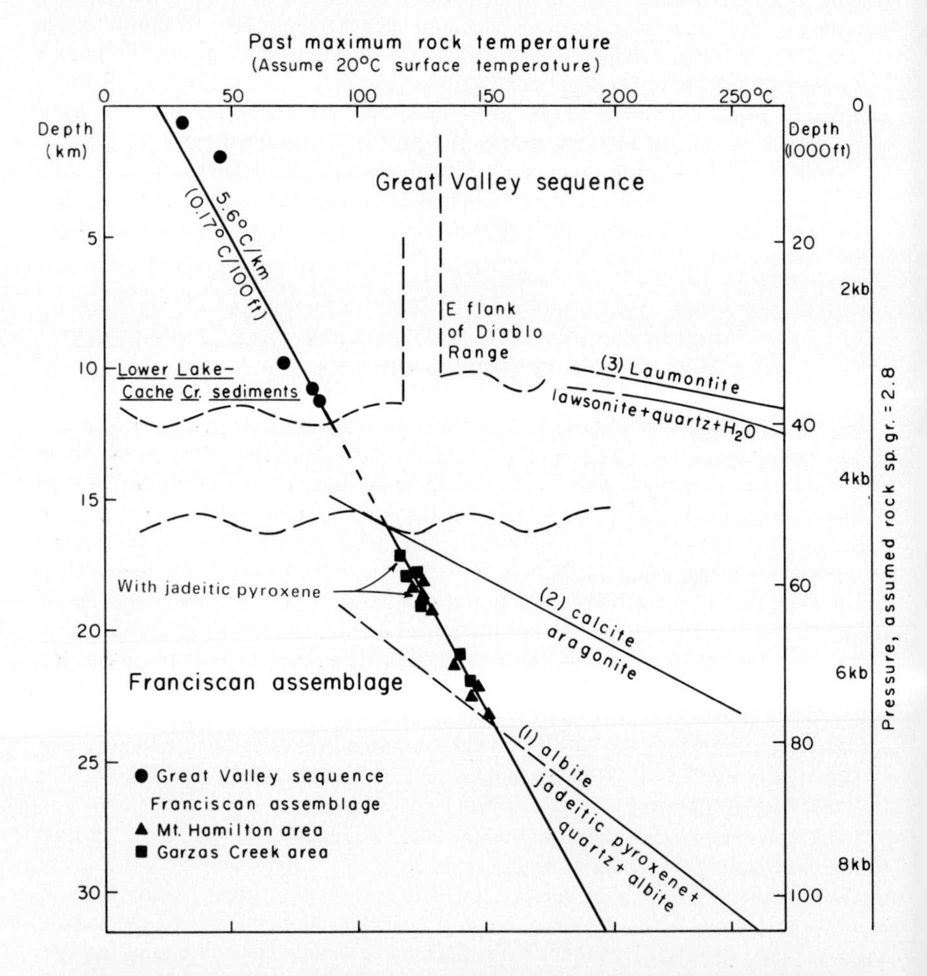


**Figure 11. Temperature-depth relations from phytoclast reflectance analysis of samples from the Great Valley sequence and Franciscan assemblage. The thermal gradient line (5.6°C per km) is based on the Great Valley sequence samples (see Table 1), and points for the Franciscan samples are plotted on this line on the basis of derived temperature only. Mineral data from: (1) Newton and Smith (1967); (2) Boettcher and Wyllie (1968); (3) Nitsch (1968); and Liou (1971). Wavy dashed lines indicate the base of the Lower Lake-Cache Creek section and the estimated former depth of the highest preserved Franciscan strata.**

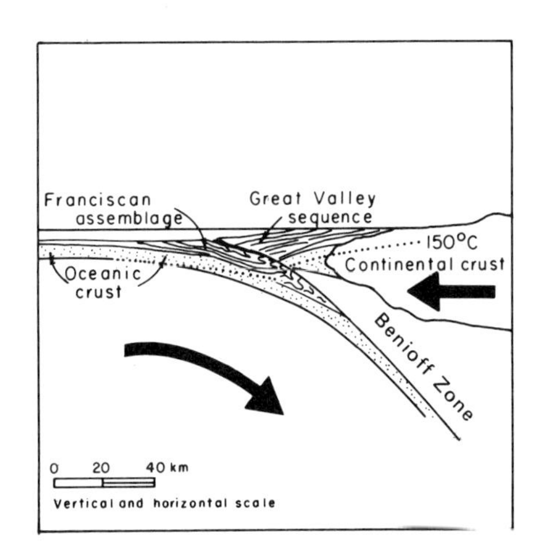


**Figure 12. Tectonic model for the relation between the Great Valley sequence and Franciscan assemblage (after Ernst, 1970, p. 894). No vertical exaggeration. Note the depression of the 150°C isotherm.**

it might be extensively found were it not that all forms of $CaCO_3$ are sparse in these rocks (Ernst, 1970).

A model that ties together the temperature data from the Diablo Range Franciscan rocks and from the Great Valley sequence was described by Ernst (1970) and is shown in Figure 12. The ocean basin, in which Franciscan rocks were being deposited, moved toward the continental margin, near which Great Valley sequence strata were accumulating; the ocean crust moved obliquely downward underneath the continental margin, perhaps forming a late Mesozoic trench, and oceanic sediments were pushed under sediments closer to the continent. This model satisfies much of the information about structure and mineral paragenesis of the terrain and, most important for this discussion, it includes depression of isotherms in the sedimentary sequence to much below normal depth. Despite variations in the model as presented by different authors, some features are persistent and agree well with the data from phytoclast metamorphism: the low thermal gradient zone is about 30 to 60 km wide, the gradient may be as low as one-third of normal even without rapid sedimentation, and the time lag of isotherm displacement is large (Minear and Toksöz, 1970, p. 1411).

To fit this model and the results of the present study of organic metamorphism, I suggest that the Great Valley sequence and Franciscan rocks now associated with them were at the same areal position in a belt of depressed thermal gradient, and I conclude that the small thermal gradient persisted at least until the end of the Eocene, when the sedimentary rocks were rapidly uplifted and eroded.

## ACKNOWLEDGMENTS

For encouragement and help during my research, I am particularly grateful to William R. Evitt, Stanford University; Win Swe, Arts and Sciences University, Mandalay, Burma; Russell R. Dutcher, Southern Illinois University; John F. Grayson, Amoco Production Company; John R. Castaño, Shell Oil Company; and M. Clark Blake, Jr., U.S. Geological Survey.

## REFERENCES CITED

Bailey, E. H., Irwin, W. P., and Jones, D. L., 1964, Franciscan and related rocks, and their significance in the geology of western California: California Div. Mines and Geology Bull. 183, 177 p.

Boettcher, A. L., and Wyllie, P. J., 1968, The calcite-aragonite transition measured in the system $CaO-CO_2-H_2O$: Jour. Geology, v. 76, p. 314-330.

Bostick, N. H., 1970, Thermal alteration of clastic organic particles (phytoclasts) as an indicator of contact and burial metamorphism in sedimentary rocks [Ph.D. dissert.]: Stanford, Calif., Stanford Univ., 220 p.

——1971, Thermal alteration of clastic organic particles as an indicator of contact and burial metamorphism in sedimentary rocks: Baton Rouge, Louisiana State Univ., Geoscience and Man, v. 3, p. 83-92.

Brown, H. R., and Taylor, G. H., 1961, Some remarkable Antarctic coals: Fuel, v. 40, no. 3, p. 211-244.

Chandra, D., and Bond, R. L., 1956, The reflectance of carbonized coals: Liege, Proc. Internat. Comm. Coal Petrology, 2d Mtg., p. 47-51.

de Vries, H.A.W., Habets, P. J., and Bokhoven, C., 1968, Das Reflexionsvermögen von Steinkohle: Brennstoff-Chemie, v. 49, p. 15-21, 47-52, 105-110.

Dickinson, W. R., Ojakangas, R. W., and Stewart, R. J., 1969, Burial metamorphism of the late Mesozoic Great Valley sequence, Cache Creek, California: Geol. Soc. America Bull., v. 80, no. 3, p. 519-526.

Ernst, W. G., 1970, Tectonic contact between the Franciscan melange and the Great Valley sequence—Crustal expression of a late Mesozoic Benioff zone: Jour. Geophys. Research, v. 75, no. 5, p. 886-901.

Forsman, J. P., and Hunt, J. M., 1958, Insoluble organic matter (kerogen) in sedimentary rocks: Geochim. et Cosmochim. Acta, v. 15, p. 170-182.

Ghosh, T. K., 1967, A study of temperature conditions at igneous contacts with certain Permian coals of India: Econ. Geology, v. 62, no. 1, p. 109-117.

Griffin, G. M., Tedrick, J. M., and Reel, D. A., 1969, Geothermal gradients in Florida and southern Georgia: Gulf Coast Assoc. Geol. Socs. Trans., v. 19, p. 189-193.

Hryckowian, E., Dutcher, R. R., and Dachille, F., 1967, Experimental studies of anthracite coals at high pressures and temperatures: Econ. Geology, v. 62, no. 4, p. 517-539.

Jaeger, J. C., 1959, The temperatures outside a cooling intrusive sheet: Am. Jour. Sci., v. 257, p. 44-54.

Kisch, H. J., 1966, Carbonization of semi-anthracite vitrinite by an analcine basanite sill: Econ. Geology, v. 61, no. 6, p. 1043-1063.

Liou, J. G., 1971, P-T stabilities of laumontite, wairakite, lawsonite, and related minerals in the system $CaO-Al_2O_3-2SiO_2-H_2O$: Jour. Petrology, v. 12, no. 2, p. 379-411.

Lovering, T. S., 1935, Theory of heat conduction applied to geologic problems: Geol. Soc. America Bull., v. 46, p. 69-94.

Minear, J. W., and Toksöz, M. N., 1970, Thermal regime of a downgoing slab and new global tectonics: Jour. Geophys. Research, v. 75, no. 8, p. 1397-1419.

Moses, P. L., 1961, Geothermal gradients now known in greater detail: World Oil, v. 152, no. 6, p. 79-82.

Newton, R. C., and Smith, J. V., 1967, Investigations concerning the breakdown of albite at depth in the earth: Jour. Geology, v. 75, p. 268-286.

Nitsch, K. H., 1968, Die Stabilität von Lawsonit: Naturwissenschaften, v. 55, p. 388.

Ronov, A. B., 1958, Organic carbon in sedimentary rocks (in relation to the presence of petroleum): Geochemistry, no. 5, p. 510-536.

Schrayer, G. J., and Zarella, W. M., 1966, Organic chemistry of shales—Pt. 2, Distribution of extractable organic matter in siliceous Mowry shale, Wyoming: Geochim. et Cosmochim. Acta, v. 30, p. 415-434.

Soliman, S. M., 1965, Geology of the east half of the Mt. Hamilton quadrangle, California: California Div. Mines and Geology Bull. 185, 32 p.

Stevenson, D. L., and Dickerson, D. R., 1969, Organic geochemistry of the New Albany Shale in Illinois: Illinois Geol. Survey Illinois Petroleum 90, p. 3-10.

Swe, Win, and Dickinson, W. R., 1970, Sedimentation and thrusting of late Mesozoic rocks in the Coast Ranges near Clear Lake, California: Geol. Soc. America Bull., v. 81, no. 1, p. 165-188.

Trask, P. D., and Patnode, H. W., 1942, Source beds of petroleum: Am. Assoc. Petroleum Geologists, 566 p.

Wang, Y., 1963, Geology and coal metamorphism by dacite in the Chiufen gold mine, northern Taiwan [M.S. dissert.]: State College, Pennsylvania State Univ., 113 p.

SYMPOSIUM HELD AT G.S.A. ANNUAL MEETING IN MILWAUKEE, NOVEMBER 1971
MANUSCRIPT RECEIVED BY THE SOCIETY MARCH 9, 1973

Printed in U.S.A.

Geological Society of America
Special Paper 153


# Microscopic Examination of Kerogen (Dispersed Organic Matter) in Petroleum Exploration

JACK D. BURGESS

*Gulf Research & Development Company*
*Houston Technical Services Center*
*Houston, Texas 77036*

## ABSTRACT

The thermal rock history of any area is manifested in the color of the recovered kerogen, or dispersed organic matter, including spores and pollen. Thermally induced color differences of kerogen viewed in transmitted light can be directly related to oil, gas, and areas barren of commerical hydrocarbon accumulations. The same technique is useful in outlining areas favorable for oil and gas exploration.

Laboratory heating experiments provided useful color standards, which, when compared with fossil residues, permit ranking kerogen alteration on a 1 to 5 thermal index scale. Subsidiary evidence that alteration and devolatilization has taken place in the darker colored organic residues is their higher reflectivity. Hydrogenation experiments by the U.S. Bureau of Mines have revealed chemical differences between the structured and amorphous type of kerogen, as well as a decrease in solubility as coal rank increases.

Utilization of kerogen color change reveals that the economic basement below which no commercial oil or gas might be found is yet unknown on the Gulf Coast.

The largest gas accumulations within the Arkoma Basin of the midcontinent are in areas of strong thermal alteration as determined by kerogen analysis of surface shales and subsurface coal samples.

## INTRODUCTION

Microscopic recognition of color changes reflecting thermal alteration in kerogen residues can be used in petroleum exploration. Morphologically distinct organic residues recovered from clastic sediments can provide clues to the likelihood of finding gas, oil, or no hydrocarbons within a region. Two separate but related lines of inquiry were pursued in this investigation:

1. The microscopic recognition of subtle differences in kerogen color as indicative of the past thermal history of the host rock was considered. These color changes are related to the maturation or thermal alteration stage of the hydrocarbon generating cycle and determine if the cycle has started, is continuing, or has gone to completion (Gutjahr, 1966).

2. Recognition and identification of morphologically distinct kerogen material making up the dispersed organic particles recovered from a rock were attempted, as well as determination of which specific end-member kerogen type can be further identified chemically in the laboratory as being a precursor of gaseous or oily hydrocarbons or unrelated to them (Breger and Brown, 1962).

The purpose, in short, is the microscopic identification of a hydrocarbon source rock by determining the thermal maturation of the recovered kerogen and the further identification of the group to which the kerogen belongs. Both of these parameters are related to the ultimate type of hydrocarbon generated. The conclusion that darkening of organic matter results from thermal alteration and that these recognized changes are related to hydrocarbon possibilities of associated sedimentary rocks is in agreement with studies of Correia (1967, 1969), Ammosov and Gorshkov (1966), Kontorovich and others (1967), and Hacquebard and Donaldson (1970).

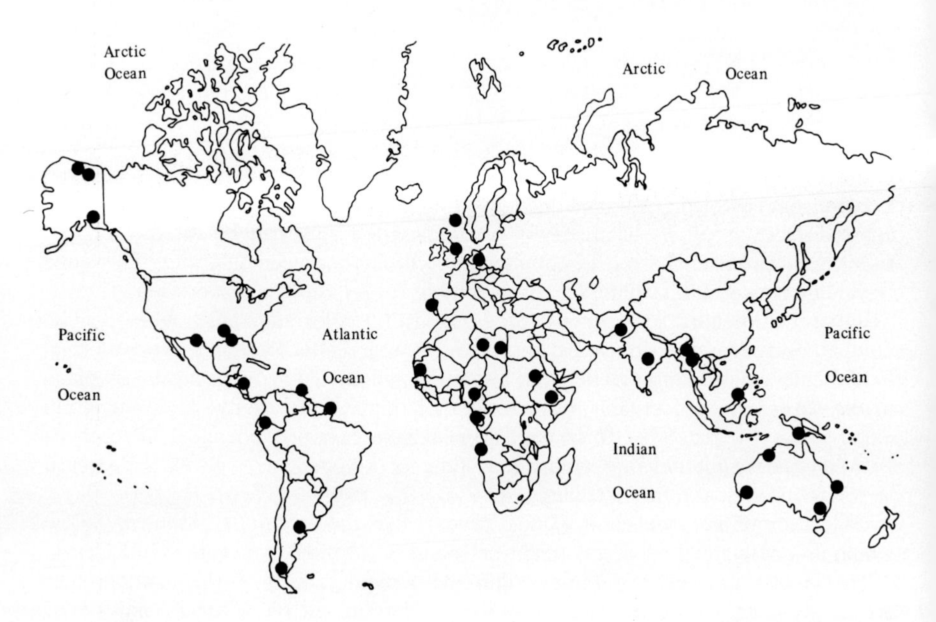


**Figure 1. Kerogen sample coverage in this study. Total number of samples, 4,600.**

## SOURCE ROCKS

What is a source rock? There is little unanimity on a standard definition, but the glossary of the American Geological Institute defines a *source rock* as "the geological formation in which oil, gas, and/or other minerals originate." *Source beds* are defined as "rocks in which oil or gas has been generated."

Keeping these definitions in mind, are there any obvious physical characteristics that can be described which set a source rock apart from a nonsource rock? It appears that there are, and the clue is in the microscopic identification of the original types of organic matter, or kerogen, mixed with mineral components during initial sedimentation. Postdepositional effects of compaction, fluid expulsion, devolatilization, and increased thermal effects resulting from basin subsidence then modify the entrapped particulate organic detritus (Teichmüller and Teichmüller, 1966). This process is referred to by Landes (1967) as eometamorphism.

The present study is based on examination of over 4,600 samples ranging in age from Ordovician to Holocene and gathered from many parts of the world, shown in Figure 1.

## KEROGEN USAGE

Kerogen, as originally defined in 1912 by Crum Brown, is the solid bituminous mineraloid substance in oil shales which yields oil upon destructive distillation. Since the term was initially introduced, it has been broadened by some geochemists and others to include all the insoluble organic matter recovered from shales and other clastic sediments after they are treated with hydrochloric and hydrofluoric acids. Kerogen, in this report, is used in the broadened context, and includes all finely disseminated organic material freed from a sedimentary rock after acid treatment. Identifiable spores, pollen, plant cuticle, wood, algae, dinoflagellates, and hystrichosphaerids, as well as a host of unidentifiable organic debris, are included in this definition. Kerogen is regarded as a complex organic mixture analogous to coal and consisting of a great variety of discréte organic particles of which only a small percentage can usually be identified microscopically.

Kerogen color used in estimating the thermal history of the host rock can be affected by processing oxidation, so this step is omitted. Slight modification of standard palynological processing, as suggested by Staplin (1969), allows for complete kerogen recovery without affecting its color.

The physical appearance of kerogen in hand specimen following acid treatment is that of a fine soft powder, varying in color from light to dark brown or black. Carbon, hydrogen, oxygen, nitrogen, and sulfur are the major chemical constituents of kerogen reported by McIver (1967). Most organic geochemists regard kerogen as a high molecular-weight compound formed, possibly by polymerization, from mobile organic substances in the rock; kerogen is regarded as intermediate between fresh organic matter on the one end, and oil or hydrocarbons on the other, as shown in Figure 2.

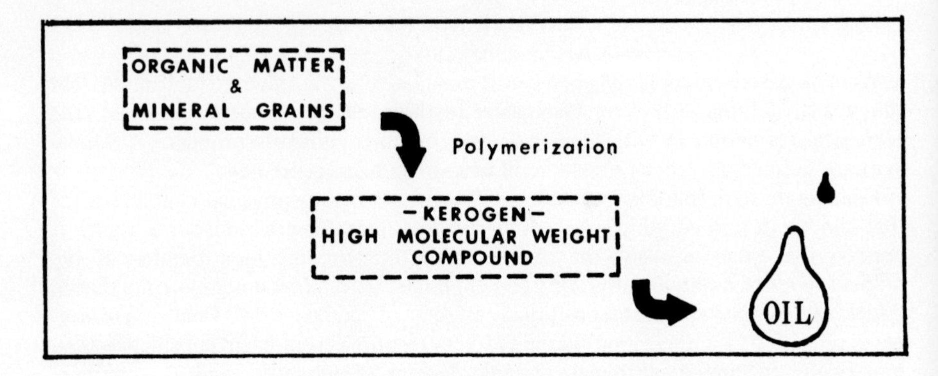


**Figure 2. Classical geochemical concept of kerogen in oil genesis.**

## THERMAL CHANGES IN KEROGEN

An increase in coal rank, recorded by either higher fixed carbon or a decrease in volatile matter with stratigraphic depth, was first recorded by Carl Hilt in 1873. This truism is referred to as Hilt's Law, and its proof follows that low-rank lignites or brown coals are never found beneath bituminous, or higher rank, coals in a normal stratigraphic sequence. Restatement and refinement of Hilt's observations as applied to petroleum exploration were set forth by White (1915 and 1935) in the now classic carbon ratio theory, which states that the past thermal history as well as the character of nearby reservoired hydrocarbons of a region can be determined by the fixed carbon values of associated coals. When coals are not present in a region, another temperature-sensitive indicator is the dispersed kerogen recovered from the sedimentary rocks.

The color of unaltered organic matter viewed through a microscope in transmitted light may be light yellow-green, orange, brown, or black, depending on the portion or type of plant which contributed the dispersed organic particle (Combaz, 1964). Because of the wide range of original colors possible in thermally unaltered organic matter, it is necessary to restrict the kerogen color evaluation to plant particles that are initially yellow, yellow green, pale orange, and light brown. The color designation is made on spores, pollen, plant cuticle, algae, and amorphous organic matter, all of which are originally in the yellow to light-brown part of the spectrum. The black fusain and vitrain common in organic residues is recorded as coaly debris, but is not utilized in designating the residue color. A typical residue is shown in Figure 3a.

Organic matter color change in nature results from heating over an interval of geologic time. In laboratory heating experiments, thermally unaltered organic matter initially changes color at just below 400°F. The first visible thermal effect is an intensification or increased brilliance of color as the yellow-greens gradually change to light orange. As temperature increases, the colors deepen from orange to light brown and eventually to black (Stadnichenko, 1926). These temperature-induced color changes are irreversible and remain upon cooling.

The laboratory recognition of orderly color changes in organic matter with increasing temperature lends itself to numerical representation as a 1 through 5 linear scale. Number 1 is assigned to thermally unaltered yellow kerogen; 2 marks the first noticeable color change below 400°F, with 3 a midpoint alteration value. The highest value is 5, and can be recognized by the colors dark brown to black. Stage 5 is identified before any mineralogic metamorphism can be detected in

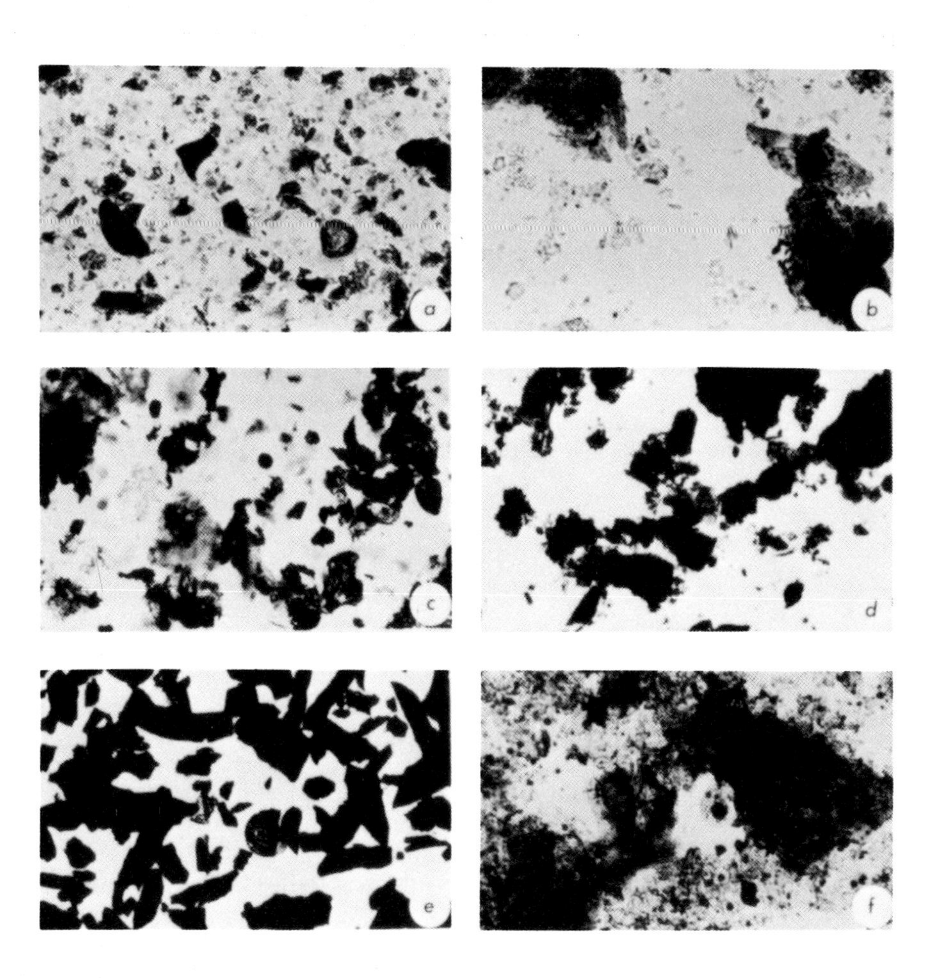


**Figure 3. All photographs are × 635. (a) Typical kerogen residue; note angular black vitrinite particles. Tertiary of Alaska. (b) Unaltered kerogen residue. The pale fragments in the background are algal debris. The larger fragments are algal masses. Recent material. (c) Kerogen residue, thermal index 3. Plant material has altered to medium brown at this stage. (d) Kerogen residue, thermal index 5; note the beginning of fragmentation as a result of thermal alteration of the thin debris adjacent to the black prominent particles. The residue is black with little or no color visible. (e) Structured plant debris residue consisting largely of fusain and vitrain. Note the single spore near the center. Pennsylvanian of Oklahoma. (f) Amorphous organic matter. This material is derived from noncalcareous colonial algae. Note granular texture of relict cell structure around periphery of the right-of-center fragment. Green River Formation.**

the rock. Examples of kerogen residues representing various thermal index stages as well as kerogen types are shown in Figure 3. The laboratory controlled standard color slides prepared from heated residues can be quickly matched with naturally occurring analogs from the fossil record.

Thermal color changes in organic matter are a visible manifestation of attendant chemical changes including devolatilization. These chemical reactions are in the direction of increased carbon percentage with a decrease in the volatile hydrogen and oxygen percentage. Sediment response to increased temperature and overburden pressure resulting from basin subsidence causes sediment compaction leading to migration of the generated volatiles from the organic matter. This process is continuous from initial living matter to graphitization, but at differing rates dependent upon the chemical composition of the original organic matter (Schopf, 1948). The sequential darkening of the organic matter is usually accompanied by increasing sediment compaction and bulk density as evidenced by increased hardness of the shales and resultant pore-size reduction.

Organic matter color, relative amount of kerogen type present, state of preservation of the organic matter, and thermal index number based on the kerogen color are recorded on a data sheet as each residue sample is microscopically examined. An example of a data sheet with the method of plotting kerogen characteristics of a hypothetical sample is shown in Figure 4. These data, thus recorded, can be used in subsequent mapping, revealing regional differences in the thermal alteration index of stratigraphic intervals of interest along with the various types of identified kerogen.

## TYPES OF ORGANIC MATTER

A second but no less important factor in microscopic source-rock recognition is the dominant type of organic matter recovered from the rock.

Presently, two broad categories of organic matter, or kerogen, can be microscopically recognized: structured plant debris and amorphous debris.

Structured plant debris is derived principally from higher land plants and consists of discrete organic particles, usually brown to black in color, with recognizable morphology as shown in Figure 3e. Chemically, this plant debris consists of cellulose, lignin, carbohydrates, proteins, and waxes, which normally produces peat or coal. Degradation of the celluloses by either diagenesis or thermal alteration may produce methane.

Amorphous debris consists of relatively structureless masses of organic matter, often with granular texture, but usually without other distinctive structure and in the thermally unaltered state bright yellow to orange in color. The amorphous material illustrated in Figure 3f originated in an aqueous environment as noncalcareous algae, but only rarely is distinctive algal structure preserved following diagenesis and compaction. This amorphous debris is rich in the "lipid fraction," which in a broad sense is closer to liquid hydrocarbons than are the celluloses of the structured plant debris mentioned above. All organic matter examined to date from worldwide oil shales is of the amorphous type.

Forsman and Hunt (1958) chemically recognized two types of kerogen, one type

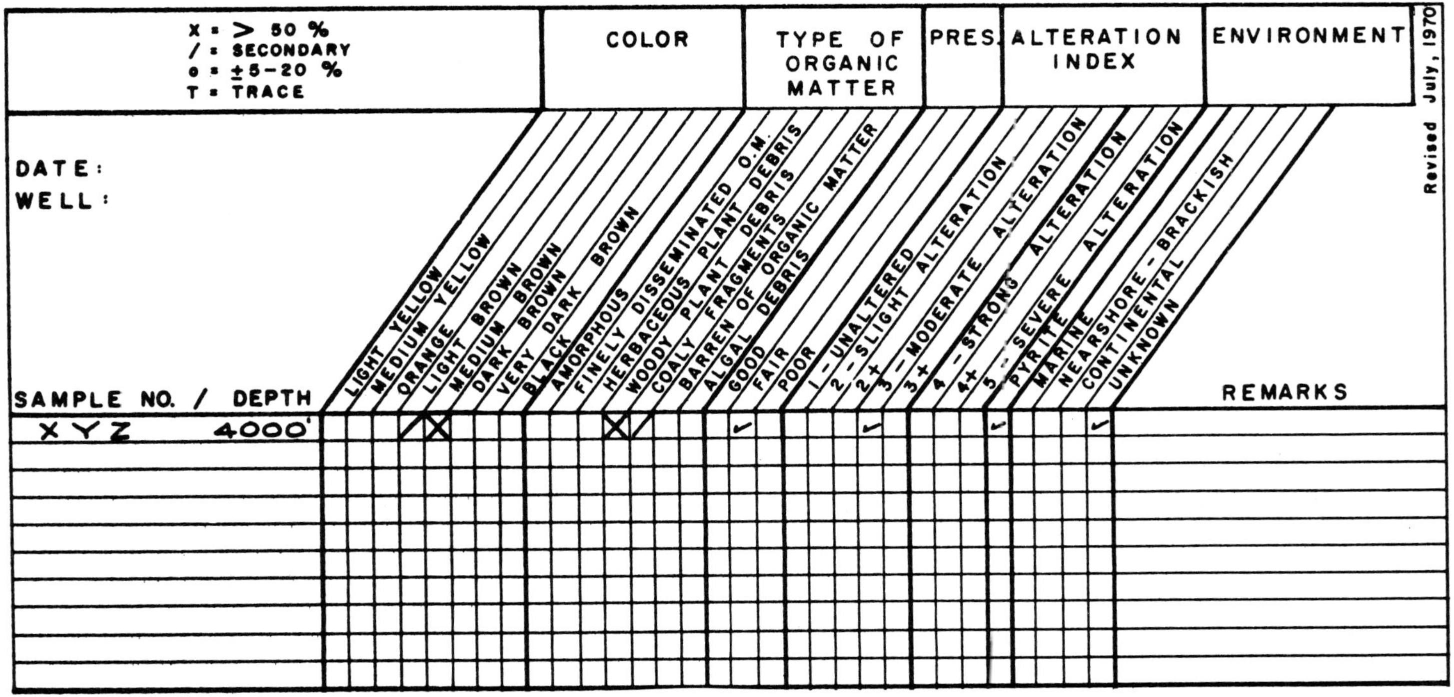

X = > 50 %
/ = SECONDARY
o = ±5–20 %
T = TRACE

DATE:
WELL:

| SAMPLE NO. / DEPTH | COLOR | | | | | | | | TYPE OF ORGANIC MATTER | | | | | | | PRES. | | | ALTERATION INDEX | | | | | | | | | ENVIRONMENT | | | | REMARKS |
|---|---|---|---|---|---|---|---|---|---|---|---|---|---|---|---|---|---|---|---|---|---|---|---|---|---|---|---|---|---|---|---|---|
| | LIGHT YELLOW | MEDIUM YELLOW | ORANGE | LIGHT BROWN | MEDIUM BROWN | DARK BROWN | VERY DARK BROWN | BLACK | AMORPHOUS | FINELY DISSEMINATED O.M. | HERBACEOUS PLANT DEBRIS | WOODY PLANT DEBRIS | COALY FRAGMENTS | BARREN OF ORGANIC MATTER | ALGAL DEBRIS | GOOD | FAIR | POOR | 1 – UNALTERED | 2 – SLIGHT ALTERATION | 2+ | 3 – MODERATE ALTERATION | 3+ | 4 – STRONG ALTERATION | 4+ | 5 – SEVERE ALTERATION | PYRITE | MARINE | NEARSHORE | CONTINENTAL - BRACKISH | UNKNOWN | |
| XYZ 4000' | | | | / | X | | | | | | | X | / | | | | ✓ | | | | | ✓ | | | | | ✓ | | | | ✓ | |

Revised July, 1970

Figure 4. An example of recording a kerogen sample observation. Sample XYZ from 4,000 ft. Sample color is medium to light brown and consists of woody plant debris with secondary amounts of coaly fragments in a fair state of preservation. The sample is altered to stage 3 based on kerogen color, contains finely disseminated pyrite, and represents an unknown environment of deposition.

like coal, the other like the kerogen from oil shales. They demonstrated quantitative differences in the solubilities of the two kerogen types following hydrogenolysis. They reported kerogen solubility of 36 percent from a bituminous coal, while solubilities of 71 and 87 percent, respectively, were achieved from two oil shale-type kerogen samples under the same laboratory conditions. This difference in solubilities of the two end-member organic matter kerogen types suggests significant chemical differences between them.

Fisher (1942), at the U.S. Bureau of Mines, showed that the hydrogen-rich constituents, spores and oil algae, were more readily liquified by hydrogenation than were the cellulosic woody coal constituents. Fisher also showed that coals of progressively higher rank produced successively less soluble material under the same laboratory conditions. These data suggest chemical differences between recognizable kerogen types, as well as significant effects of thermal alteration rank on kerogen solubility characteristics.

The interpretation of the role kerogen plays in oil genesis based on this study is shown in Figure 5.

## EXAMPLES OF KEROGEN STUDY IN OIL EXPLORATION

An empirical relation between kerogen alteration and areas productive of oil, dry gas, or no commerical hydrocarbons has been documented in western Canada by Staplin (1969) and in the Bowen Basin of Australia by Evans (1963). Areas where the thermal index of kerogen is rated 1 through 3 may give up oil or wet gas; areas rated 4 usually yield only dry gas; areas rated 5 seldom produce any but trace quantities of hydrocarbons.

Mapping organic matter alteration changes can emphasize the most favorable areas for hydrocarbon exploitation. Conversely, the same technique can be utilized to warn of severely altered areas where the probability of finding commercial

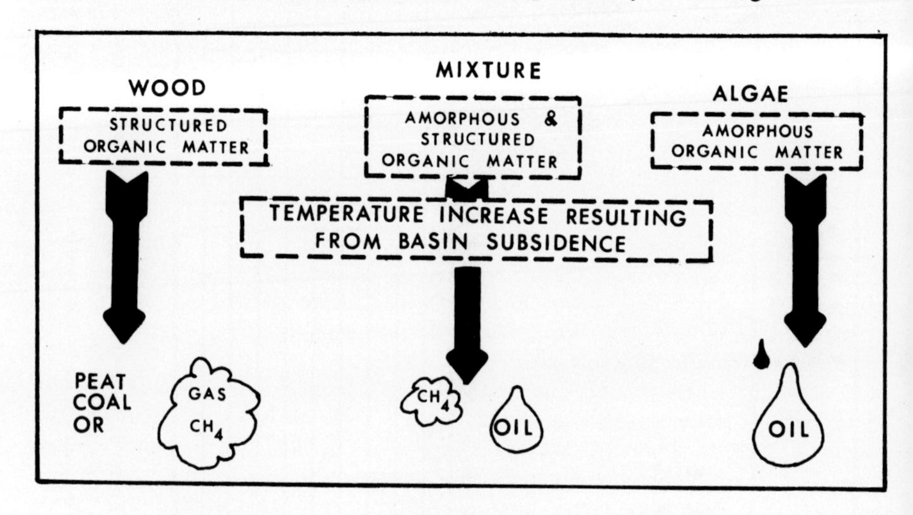


Figure 5. Role of kerogen in oil genesis, based on this study.

hydrocarbon accumulations would be minimal. Kerogen data can also be incorporated into regional geologic evaluation of hydrocarbon prospective concessions, both foreign and domestic, to determine if acquisition is economically attractive.

Economic basement has not been determined in the offshore Gulf of Mexico. Depths of from 13,800 ft downward through 18,000 ft in the offshore Gulf Coast reveal visual alteration values of from a low of 2 to a high of 3+ on our 1 to 5 scale and are well within the alteration range of liquid hydrocarbon production elsewhere in the world. Economic basement in this important oil province is at some depth in excess of the 18,000 ft thus far evaluated by kerogen alteration studies.

An isocarb map of the Atokan formation in the Arkoma Basin is shown in Figure 6. This map was constructed on the kerogen recovered from surface samples and on published coal-fixed-carbon data corrected to a dry ash-free basis (Moose and Searle, 1929). Alteration patterns in the Pennsylvanian Atokan formation generally increase from west to east, as pointed out by Wilson (1961). A lobe of slight alteration trends northwest onto the Oklahoma Platform and is at variance with the general southeastward alteration increase into the deeper portion of the basin. The greatest intensity of alteration is found in the center of the basin adjacent to the Oklahoma-Arkansas boundary, south of Fort Smith.

A relation between gas fields and alteration isocarbs is evident in Figure 6. The majority of the larger gas fields are enclosed within the 3 or higher alteration isocarb, suggesting that thermal alteration probably exerted control both in generation and trapping of the gaseous hydrocarbons found in the basin. Kerogen alteration, along with coal-fixed-carbon analysis can be used in depicting regional alteration trends. Future work in kerogen reflectance microscopy may expand this technique and bring it into more widespread use in recognizing alteration trends in petroleum exploration.

Kerogen alteration evidence compiled within a region can probably never be positive enough to condemn an otherwise favorable petroleum structural prospect. However, if the first test drilled on the favorable structure is a dry hole, confirming the unfavorable kerogen alteration interpretation, it should then preclude the need and expense of drilling other tests on this same structure.

## SUMMARY

It is possible to relate the color of dispersed organic matter or kerogen under the microscope to the past thermal history of an area or region. Utilizing this information can provide insight into the hydrocarbon possibilities of either a stratigraphic interval, area, or entire basin, as well as the probable type of hydrocarbon that can be expected.

In addition, whenever we can microscopically identify a reasonable quantity of amorphous organic matter recovered from a rock subjected to no more than moderate thermal alteration, we can confidently say that this is a good source rock. This observation has been repeatedly confirmed by geochemical source-rock analyses. The same confidence level does not exist when the structured plant debris residues, consisting largely of fusain and vitrain, are identified.

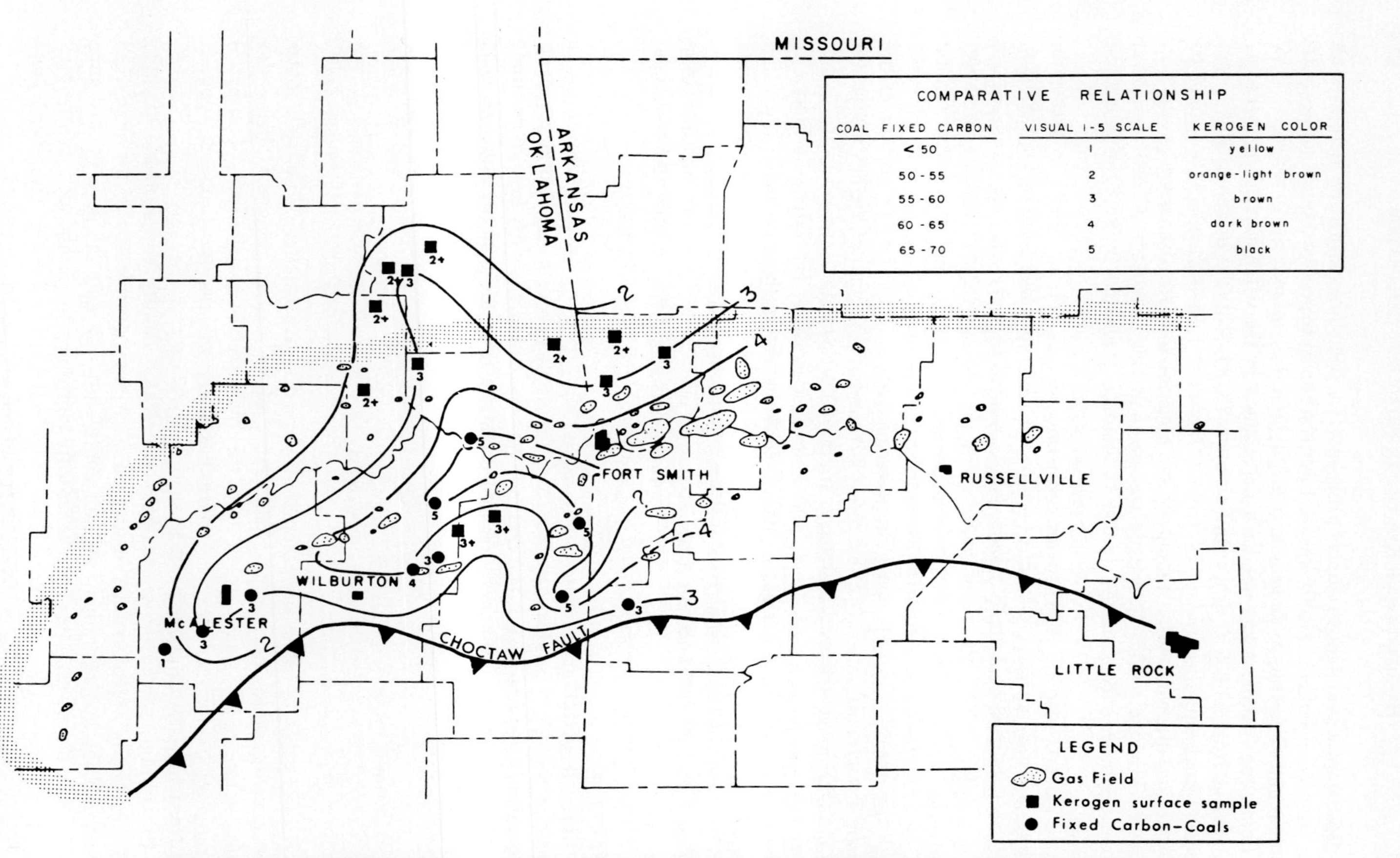


**Figure 6. Arkoma Basin thermal alteration isocarb map, interpreted from kerogen and coal-fixed-carbon data.**

The microscopic recognition of kerogen type as well as the thermal history of the rock from which it was recovered can significantly contribute to geochemical source-rock identification. In the past, organic geochemists have paid too little heed to the variety of materials making up the kerogen residue, while many paleobotanists were often unaware of the importance of specific types of organic matter to hydrocarbon generation. It now appears possible for these two disciplines to work together in solving the fascinating problem of the origin of petroleum.

## ACKNOWLEDGMENTS

I thank Esso Production Research for permitting publication of this paper. My present employer, Gulf Oil company, prepared the illustrations, for which I am most grateful.

## REFERENCES CITED

Ammosov, I. I., and Gorshkov, V. I., 1966, The katagenesis of rocks and oil and gas occurrence in the west Siberian lowland: Akad. Nauk SSSR Doklady, v. 161, no. 1, p. 159-162.

Breger, I. A., and Brown, Andrew, 1962, Kerogen in the Chattanooga Shale: Science, v. 137, no. 7, p. 221-227.

Combaz, A., 1964, Les Palynofacies: Rev. Micropaléontologie, v. 7, no. 3, p. 205-218.

Correia, M., 1967, Possible relation between the state of preservation of visible elements of organic matter (polynoplanktological microfossils) and the existence of hydrocarbon deposits: Inst. Français Pétrole Rev., v. 22, no. 9, p. 1285-1306.

——1969, Contribution à la recherche de zones favorables a la genèse du pétrole par l'observation microscopique de la matière organique figurée: Inst. Français Pétrole Rev., v. 24, no. 12, p. 1417-1454.

Evans, P. R., 1963, Spore preservation in the Bowen Basin, Australia: Australia Bur. Mineral Resources, Geology and Geophysics, 19 p., 2 maps.

Fisher, C. H., 1942, Hydrogenation and liquifaction of coal, Pt. 2: Effect of petrographic composition and rank of coal: U.S. Bur. Mines Tech. Paper 642.

Forsman, J. P., and Hunt, J. M., 1958, Insoluble organic matter (kerogen) in sedimentary rocks of marine origin, *in* Weeks, L. G., ed., Habitat of oil: Am. Assoc. Petroleum Geologists, p. 747-778.

Gutjahr, C.C.M., 1966, Carbonization measurements of pollen grains and spores and their application [Ph.D. thesis]: Leiden, Netherlands, Leiden Univ., J. J., Groen and Zoon, 29 p.

Hacquebard, P. A., and Donaldson, J. R., 1970, Coal metamorphism and hydrocarbon potential in the upper Paleozoic of the Atlantic provinces, Canada: Canadian Jour. Earth Sci., v. 9, no. 4, p. 1139-1163.

Kontorovich, A. E., Parparova, G. M., and Trushkov, P. A., 1967, Metamorphism of organic matter and certain problems of oil and gas occurrence: Akad. Nauk SSSR, Sibirskoe Otdelenia Geologiya i Geofizika, no. 2, p. 16-29.

Landes, K. K., 1967, Eometamorphism, and oil and gas in time and space: Am. Assoc. Petroleum Geologists Bull., v. 51, no. 6, pt. 1, p. 828-841.

McIver, R. D., 1967, Composition of kerogen—Clue to its role in the origin of petroleum: Mexico City, 7th World Petroleum Cong. Proc., p. 25-36.

Moose, J. E., and Searle, V. C., 1929, A chemical study of Oklahoma coals: Oklahoma Geol. Survey Bulls. 50 and 51, 112 p.

Schopf, J. M., 1948, Variable coalification: The process involved in coal formation: Econ. Geology, v. 43, no. 3, p. 207-225.
Stadnichenko, Taisia, 1926, Microthermal studies of some "mother rocks" of petroleum from Alaska: Am. Assoc. Petroleum Geologists Bull., v. 13, p. 23.
Staplin, F. L., 1969, Sedimentary organic matter, organic metamorphism, and oil and gas occurrence: Canadian Petroleum Geologists Bull., v. 17, no. 1, p. 47-66.
Teichmüller, M., and Teichmüller, R., 1966, Geological causes of coalification: Washington, D.C., Advances in Chemistry Ser. 55, Am. Chem. Soc., p. 133-155.
White, David, 1915, Some relations in origin between coal and petroleum: Washington Acad. Sci. Jour., v. 5, no. 6, p. 189-212.
——1935, Metamorphism of organic sediments and derived oils: Am. Assoc. Petroleum Geologists Bull., v. 19, no. 5, p. 589-617.
Wilson, L. R., 1961, Palynological fossil response to low-grade metamorphism in the Arkoma Basin: Tulsa Geol. Soc. Digest, v. 29, p. 131-140.

Symposium Held at G.S.A. Annual Meeting in Milwaukee, November 1971
Manuscript Received by the Society March 9, 1973
Author's Present Address: Gulf Research & Development Company, P.O. Drawer 2038, Pittsburgh, Pennsylvania 15230

Printed in U.S.A.

Geological Society of America
Special Paper 153


# Interpretation of Vitrinite Reflectance Measurements in Sedimentary Rocks and Determination of Burial History Using Vitrinite Reflectance and Authigenic Minerals

JOHN R. CASTAÑO
AND
DENNIS M. SPARKS

*Shell Oil Company*
*Los Angeles, California 90051*

## ABSTRACT

Vitrinite reflectance studies in non-coal-bearing rocks have been conducted over the past several years; the techniques for extraction, analysis, and interpretation of small vitrinite particles vary slightly from standard coal petrographic procedures.

Two areas in California afford a comparison of thermal history utilizing authigenic minerals and vitrinite reflectance: the Great Valley sequence of late Mesozoic age near Cache Creek west of the Sacramento Valley, and Miocene to Eocene sediments in the Tejon area of the San Joaquin Valley. In the Tejon area, laumontite is a postcompaction authigenic mineral and first occurs at 9,500 ft and is abundant below 10,500 ft. Vitrinite reflectance studies of Tejon area wells demonstrate an equivalent coal rank of high-volatile C bituminous to subbituminous A at 10,500 ft. At this depth, the temperature is 218°F (104°C), and the hydrostatic pressure is 4,600 psi (317 bars). Electrical log temperatures, as corrected by comparison with accurate subsurface temperature profiles, were used to reconstruct paleothermal conditions.

At Cache Creek, laumontite is found as an alteration product in sandstones formerly buried at depths greater than 32,500 ft; the calculated paleopressure at this level is 27,000 psi (1,860 bars). Vitrinite reflectance measurements within the laumontite-bearing interval indicate a thermal history equivalent to high-volatile B bituminous coal. Based on the temperature-reflectance relation at Tejon, the temperature at the top of the laumontite zone at Cache Creek is estimated to be 265°F (130°C), and the paleogeothermal gradient in the Great Valley sequence is around 0.62°F per 100 ft (1.1°C per 100 m). Because the formation of zeolites is governed by a complex interaction of physical and chemical factors, we judge that vitrinite reflectance, which is largely temperature dependent, is the more promising tool for the measurement of thermal history.

## INTRODUCTION

During the past several years, reflectance analysis of coal macerals has been utilized as an important technique in the petrographic study and evaluation of coals. Standardization of methods and procedures has led to a very useful and precise coal ranking scheme. For several years, the same principles and techniques have also been applied to macerals in noncoaly sedimentary rocks. A variety of geologic problems has been approached, but basically the purpose has been the same as that in coal studies, namely, development of a ranking parameter that is a measure of the metamorphic history of the rocks in question.

The purpose of this paper is to review the techniques used for reflectance analysis of vitrinite in noncoaly clastic rocks and review of some results from those analyses. We will present a comparison of data obtained by reflectance analysis with that obtained from a study of authigenic minerals whose occurrence may also be a clue to the metamorphic history of the rock. The samples for the latter comparison are taken from measured stratigraphic sections of Cretaceous and Jurassic rocks west of the Sacramento Valley and from subsurface intervals of Tertiary age in the Tejon oil field area of the San Joaquin Valley. Both of these sample areas had been studied previously, and published data were available on the authigenic mineral assemblages present.

## INTERPRETATION OF VITRINITE REFLECTANCE DATA IN NON-COAL-BEARING SEQUENCES

Standard ASTM (American Society for Testing Materials) procedures are used for coal specimen preparation [ASTM D 2797 (ASTM, 1970)]. For clastic rocks in which carbonaceous macerals are accessory constituents, concentration of the vitrinite is necessary. Various modifications of palynological procedures have been utilized, and experimental data (Table 1) indicate that nonoxidative acid solution of the inorganic matrix has little effect on the reflectance of vitrinite concentrated in this way. Most clastic sediments yield considerable vitrinite from a single-step, hydrofluoric-acid-insoluble residue.

Analyses are made with a Leitz Ortholux reflecting microscope equipped with a voltage stabilizer and Photovolt Model 520 photomultiplier. The measuring aperture

TABLE 1. COMPARISON OF VITRINITE REFLECTANCE DATA BETWEEN TREATED AND UNTREATED FRACTIONS OF COAL SAMPLES

| Coal rank | Mean reflectance (% in oil) | |
|---|---|---|
| | 1 | 2* |
| Lignite | 0.30 | 0.30 |
| Subbituminous | 0.40 | 0.39 |
| Subbituminous | 0.55 | 0.56 |
| High-volatile C bituminous | 0.46 | 0.46 |
| High-volatile B bituminous | 0.54 | 0.51 |
| High-volatile A bituminous | 0.70 | 0.69 |
| High-volatile A bituminous | 0.82 | 0.83 |
| Medium-volatile bituminous | 1.07 | 1.14 |
| Medium-volatile bituminous | 1.29 | 1.31 |
| Low-volatile bituminous | 1.41 | 1.41 |
| Semianthracite | 2.00 | 1.96 |

*Samples treated with hydrochloric and hydrofluoric acids.

is three to four microns. The data are recorded on a Beckman strip chart recorder and are transferred to IBM punch cards by use of a Benson-Lehner x-y coordinate reader.

Our initial efforts were directed at establishing confidence in the correlation between standard ASTM ranking parameters [ASTM 1970 D 388 (ASTM, 1970)] and vitrinite reflectance. Figure 1 shows the fitted curve of reflectance versus coal rank for some 425 samples, 55 of which were run in our own laboratory. The primary sources of published data were papers by Benedict and Berry (1966), Gray and Shapiro (1966), and de Vries and others (1968). Only analyses reporting standard ASTM rank and mean maximum reflectance were used. The correlation is good, and we feel that much of the variance observed is due to compositional variables that affect the whole coal analyses rather than variance in the reflectance values.

In order to establish an experimental basis for the validity of the reflectance calibration of vitrinite and rank for non-coal-bearing sequences, we examined a number of coal-shale pairs. There seems to be no *a priori* reason that the low-grade biogenic and thermal processes that are responsible for the formation of vitrinite in coal would differ in either quality or intensity in associated fine-grained clastic sediments. Figure 2 shows the results of the analysis of a coal-claystone pair from the Kenai Formation, Cook Inlet, Alaska. It is apparent that the mean values are from the same population, although the range of values is greater in the claystone. Many such pairs have been analyzed, and the results have been similar, both at different ranks and with varying lithologies. The variation in reflectance values from vitrinite in noncoal sediments tends to increase inversely with the absolute amount of vitrinite in the rock. This suggests that the variance is largely a statistical relic due to inadequate sample size, and several statistical models have been used to overcome this source of error. There are, however, practical problems in evaluating reflectance data from sediments where very large sample sizes are not possible (well cuttings, for example), and we have made an attempt to understand and correct some of them. The two most obvious sources of error in our analyses

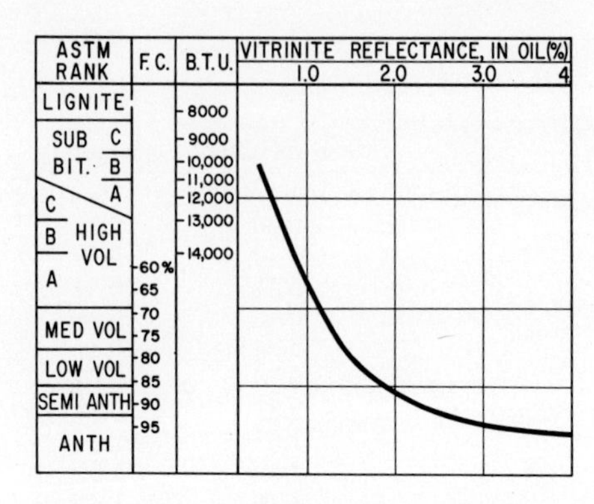


**Figure 1. Relation of vitrinite reflectance to ASTM whole-coal rank.**

have been weathering or oxidative effects and reworking of second-cycle materials. Both of these factors operate in all depositional settings, but where the absolute amount of vitrinite in the rock is small, and the particle size is also small, the influence may be relatively large, perhaps even limiting the usefulness of the method in some cases. Figure 3 depicts two superimposed histograms from cores in a well from the Cook Inlet area, Alaska. The Eocene sample is a carbonaceous claystone from the nonmarine Kenai Formation, and the Upper Cretaceous sample is a bathyal shale in a turbidite sequence. The vitrinite reflectance histogram of the Upper Cretaceous sample is strongly bimodal; this is very common in turbidites. Pollen carbonization studies by R. V. Emmons (1967, written commun.) support the conclusion that the mean reflectance is around 0.65 percent and not 1.2 to 1.4 percent. The higher reflectance vitrinite is considered to be reworked from older rocks or, perhaps, to have been partly oxidized before reaching the depositional site. As a general rule, the lowest reflectance vitrinite group in a sample is chosen as representative of the correct coal-rank equivalent.

A critical element in the evaluation of vitrinite reflectance is a knowledge of the lithology and of the depositional environment. It is well known that oxidation increases the reflectance; extreme oxidation is easily recognized, but lesser degrees of oxidation are usually hard to detect. The cause of bimodal histogram patterns can be a physical incorporation of older material, as suggested above. Bioturbation by marine organisms accompanied by repeated winnowing action, which is prevalent in shallow marine environments, can also be a factor in the preservation of organic

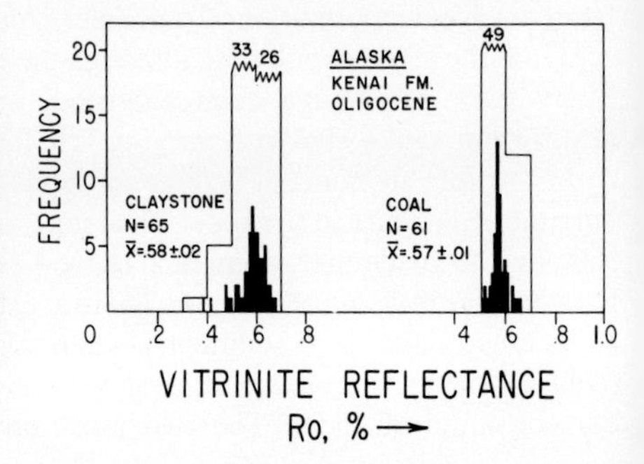


**Figure 2. Coal-claystone pair, Kenai Formation, Alaska. The reflectance data on these two histograms are grouped into classes of 0.1 percent Ro (reflectance in oil), and the individual readings are shown by vertical black bars. N refers to the number of readings used in the calculations, and $\bar{x}$ is the arithmetic mean of the mean maximum reflectance. The ± value after the $\bar{x}$ is the 95 percent confidence limits (the t distribution of Student in Dean and Dixon, 1951); the confidence limits are expressed in percent Ro. The narrower the confidence limits are, the better the sample.**

material (Figs. 4 and 5). This particular sample is judged to have a very large amount of reworked vitrinite; obvious oxidation is seen only in those particles with reflectances greater than 1.0 percent. Sandstones were found to contain a high proportion of partly oxidized vitrinite, and we generally avoid the coarse clastic rocks despite the fact that large carbonaceous fragments are often present.

## FACTORS GOVERNING COALIFICATION

Various methods have been employed by the geologist seeking to determine the burial metamorphic history caused by burial of a particular section; organic entities such as coal are very useful, as they are more sensitive than minerals to changes in temperature and are less sensitive to pressure changes (Teichmüller and Teichmüller, 1966, 1968; Kisch, 1969). Although vitrinite is not a homogeneous or clearly defined substance, vitrinite reflectance is the most suitable coal-ranking parameter for coalification studies, which, in turn, can be related to burial metamorphic mineral facies.

The conditions controlling the coalification process have been summarized by Teichmüller and Teichmüller (1966, 1968); they concluded that temperature is a critical factor in the chemical reactions involved, while the effect of pressure on the increase of coal rank is only physical, that is, compression and consolidation accompanied by losses of volume and moisture. They showed that coal rank is at a minimum on the southern margins of the Ruhr Basin where folding is the most intense; in this area, the coal-bearing strata have undergone less burial than those farther north in spite of greater tectonic deformation. Experiments by Huck and Patteisky (1964) have shown that pressure does not advance the chemical coalification process, but actually retards it. In Bostick's (1970) hydrothermal bomb experiments, no change in vitrinite reflectance was noted in different runs where the pressure was varied from 3,000 to 24,000 psi (205 to 1,650 bars) while the temperature was kept constant at 250°C (482°F).

The rate of increase of temperature with depth depends on the geothermal gradient, which is controlled by the heat flow and the thermal conductivity of the rocks. In addition, the chemical reactions involved in the coalification process are time dependent as well as temperature dependent. Hacquebard and Donaldson (1970)

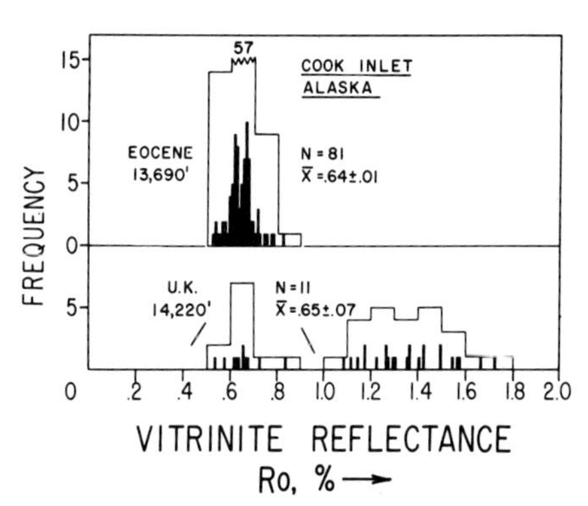


**Figure 3. Vitrinite reflectance histograms from a well in the Cook Inlet Basin, Alaska. Note bimodal distribution for U.K. sample; statistics shown are for the portion enclosed by diagonal lines.**

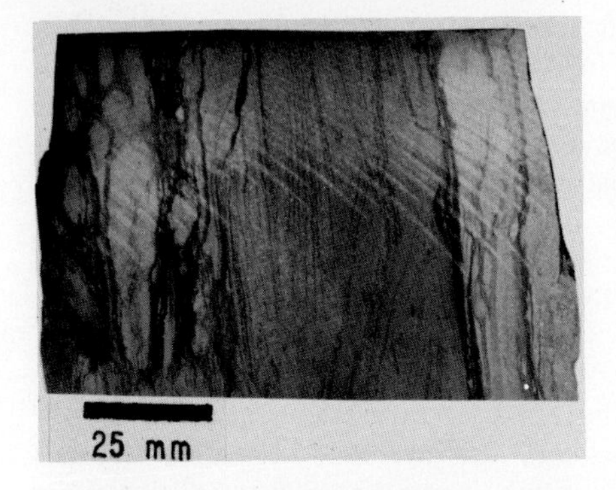


Figure 4. Photograph of slabbed core, Lower Cretaceous, Arctic Slope area, Alaska. Note cross-bedding and bioturbation. Refer also to reflectance histogram, Figure 5.

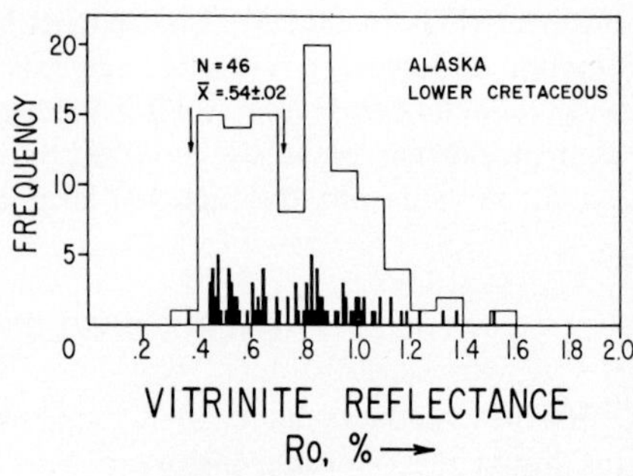


Figure 5. Histogram of a Lower Cretaceous shale, Arctic Slope, Alaska; vitrinite obtained from core sample shown in Figure 4. Statistics presented are for the part of the vitrinite population that is judged to be unaltered and is enclosed by the arrows.

compared the burial history of two lower Carboniferous coals, and using Karweil's (1956) chart (Fig. 6), they found that the 300-m.y.-old Moscow, U.S.S.R., brown coal had never been exposed to temperatures higher than 24°C (75°F), while the Sidney, Nova Scotia, high volitile A coal of the same age had been subjected to temperatues of 65°C (149°F).

In addition to geologic age and geothermal gradient, burial path and duration of the maximum heating phase are also important factors in the determination of the paleotemperature. Complications may arise in areas where subsidence has been interrupted by episodes of erosion or uplift.

Pore fluid chemistry does not appear to have any effect on coal rank and, unlike retrogressive mineral reactions the coalification process is not reversible.

## DETERMINATION OF THERMAL HISTORY USING AUTHIGENIC MINERALS AND VITRINITE REFLECTANCE

### Introduction

Two areas in California afford a comparison of thermal history utilizing authigenic minerals and vitrinite reflectance: the Great Valley sequence of late Mesozoic age on the west side of the Great Valley, and the Tejon area, which contains Miocene to Eocene rocks, at the south end of the Great Valley in the San Joaquin Basin.

### Tejon Area, California

**Introduction.** In the Tejon area (Fig. 7), there are a number of oil fields productive from sandstone reservoirs ranging in age from Miocene to Eocene. The samples were taken from the wells indicated on the map; only core samples were examined.

Present-day well depths were used in this area because, although the older beds of lower Miocene to Eocene age at the North Tejon oil field are strongly faulted and folded, the middle Miocene and younger beds exhibit a gently northerly dipping homocline (Park, 1961). It is likely that the maximum burial depth for these strata has not been much greater than it is at the present time.

**Vitrinite Reflectance Study.** The reflectance-depth plot shown on Figure 8 is a composite well profile for the Tejon area. The average line through these data is weighted heavily toward the good-quality samples. Figures 9, 10, and 11 show some of the individual reflectance histograms for the Tejon area and help illustrate the basis for the interpretation. Figure 9 shows an excellent unimodal histogram with a very small spread in reflectance. Figure 10 also exhibits an excellent unimodal pattern, and this sample is well within the laumontite-bearing interval in this area. Three superposed histograms from Richfield Kern County Land Company well B-1 are presented on Figure 11; the sample from 10,715 to 10,717 ft has very little vitrinite, some of which is probably weathered, reworked, or both, and the

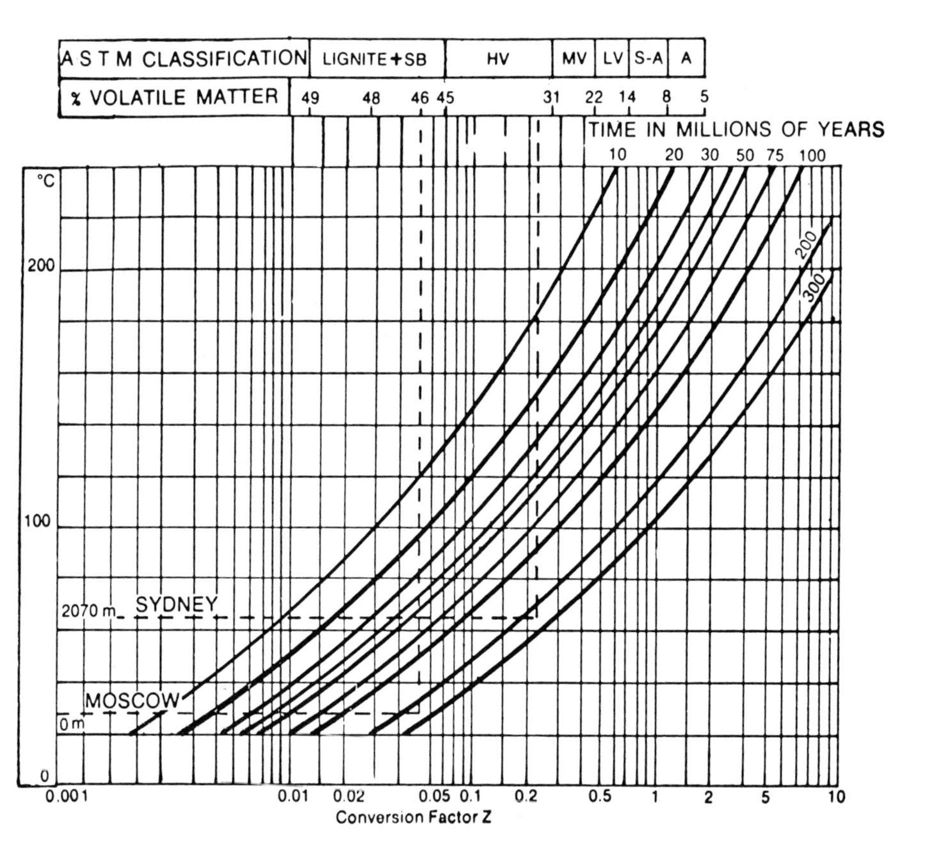


**Figure 6. Relation between rank, temperature, and time of coalification (after Karweil, 1956, as modified by Hacquebard and Donaldson, 1970). Refer to Figure 1 for coal rank-reflectance calibration.**

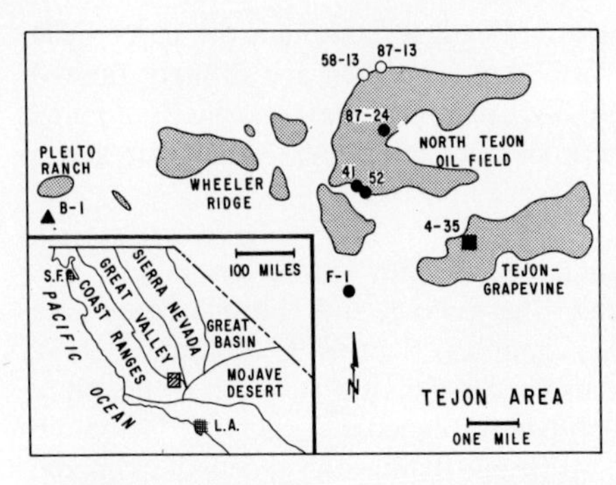


**Figure 7. Location map, Tejon area, California. The various symbols and abbreviations identify the wells examined in this study; the well symbols are also used in Figure 8. Wells labeled "41" and "52" at North Tejon oil field are referred to elsewhere as F-52-36 and F-41-36. Stippled areas are oil fields.**

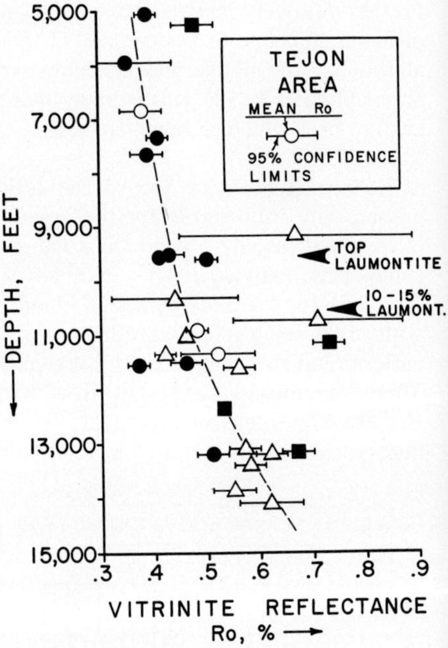

**Figure 8. Reflectance-depth plot, Tejon area, California. The symbols used in the plot identify the wells studied; refer to Figure 7 for explanation.**

results are not usable. The other two cores yielded vitrinite histograms of good to excellent quality. Returning to the composite well-reflectance profile presented on Figure 8, it is seen that a rather accurate reflectance line can be drawn by utilizing information on sample quality and particularly from a sequence of samples, although there is scatter in the mean reflectance data.

**Development of Authigenic Minerals.** Kaley and Hanson (1955) first reported the presence of laumontite and leonhardite in the Tejon area, in Standard CCMO well 4-35 at a depth of of 11,000 ft. They used microscopic and x-ray powder diffraction methods for indentification of these zeolites.

In this study, Miocene-Eocene sandstones were examined to establish the sequence of diagenetic mineralization. It was found that laumontite (and its alteration product, leonhardite) was rather easily identified in thin section; x-ray diffraction analysis corroborated the microscopic identification. The presence of the zeolites was also detected in grain-size analyses of the sandstones, for these minerals leave a silica gel residue when treated with hydrochloric acid. X-ray diffraction patterns of bulk rock samples were inadequate for identification even when the concentration was 10 to 12 percent (Fig. 12a). Where 15 to 20 percent laumontite is present, laumontite peaks appear to show up on a diffraction pattern of a bulk-rock sample (Fig. 12b). In almost pure laumontite, that is, in a situation where laumontite has replaced the carbonate in a mollusc shell, the laumontite peaks are very well displayed (Fig. 12c). Not all the sandstones are zeolitized, as much of the oil production in this immediate area is from porous, uncemented sandstones within the zone of alteration.

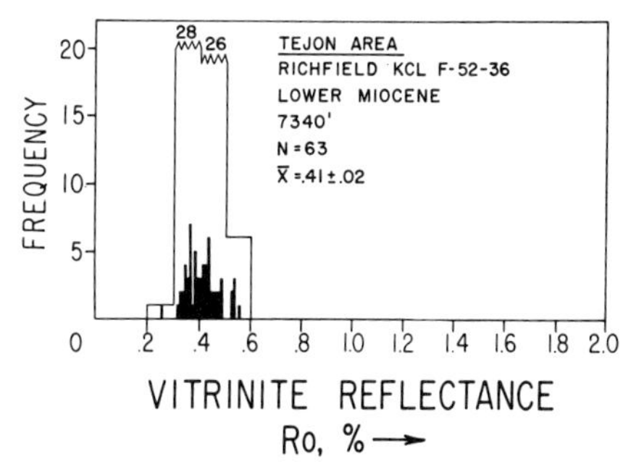


Figure 9. Vitrinite reflectance histogram, Tejon area, California. Above the laumontite-bearing zone.

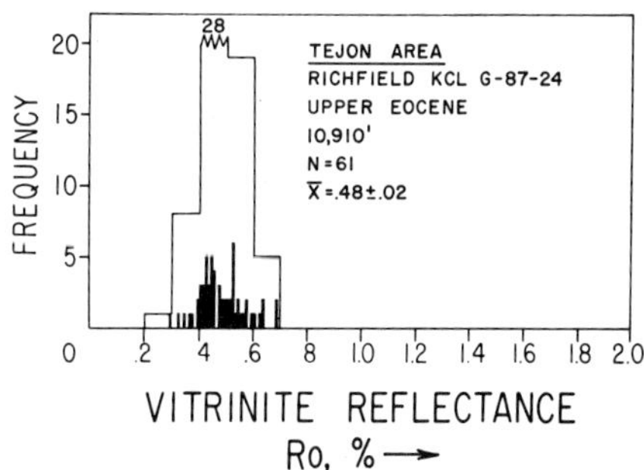


Figure 10. Vitrinite reflectance histogram, Tejon area, California. Within the laumontite-bearing interval.

Figure 13 shows a thin section of a very porous, unaltered arkosic sandstone at 5,238 to 5,263 ft; plagioclase feldspars here have a composition of $An_2$ to $An_{40}$; they typify the beds above the zeolitized section. Analcite-bearing submarine basalts of lower Miocene age are interbedded with marine sandstones and shales in the interval 7,500 to 8,500 ft. The outflow of these basalts has apparently not affected the history of mineral alteration in the subjacent section, and no anomaly in vitrinite reflectance has been observed. Laumontite is first noted at approximately 9,500 ft and is found in considerable abundance below 10,000 ft (Fig. 14). Within the zeolitized zone, plagioclase feldspars have the composition $An_2$ to $An_{15}$, although sandstones with early carbonate cement have the same composition as the sandstones above ($An_2$ to $An_{40}$). Albitization of plagioclase feldspar occurs at depths only a few hundred feet shallower in the wells of the Tejon area than does laumontization.

**Subsurface Temperatures and Calculation of Geothermal Gradient.** Accurate bottom-hole temperature data are extremely useful in relating the changes that take place in the organic and inorganic matter within a sedimentary basin, beginning with early diagenesis and continuing into the domain of low-grade metamorphism.

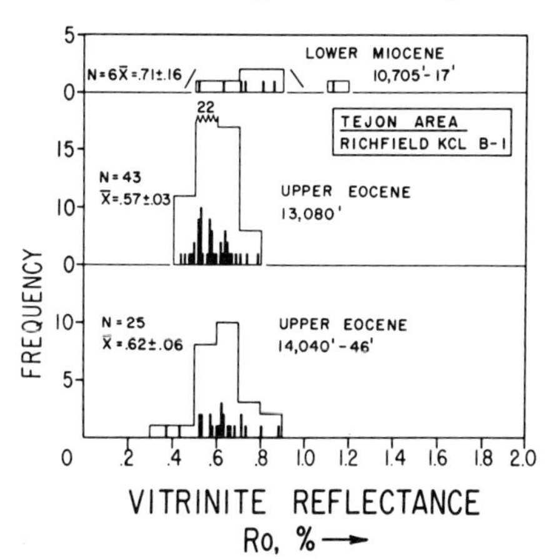


Figure 11. Vitrinite reflectance histograms, Richfield KCL B-1, Tejon area, California.

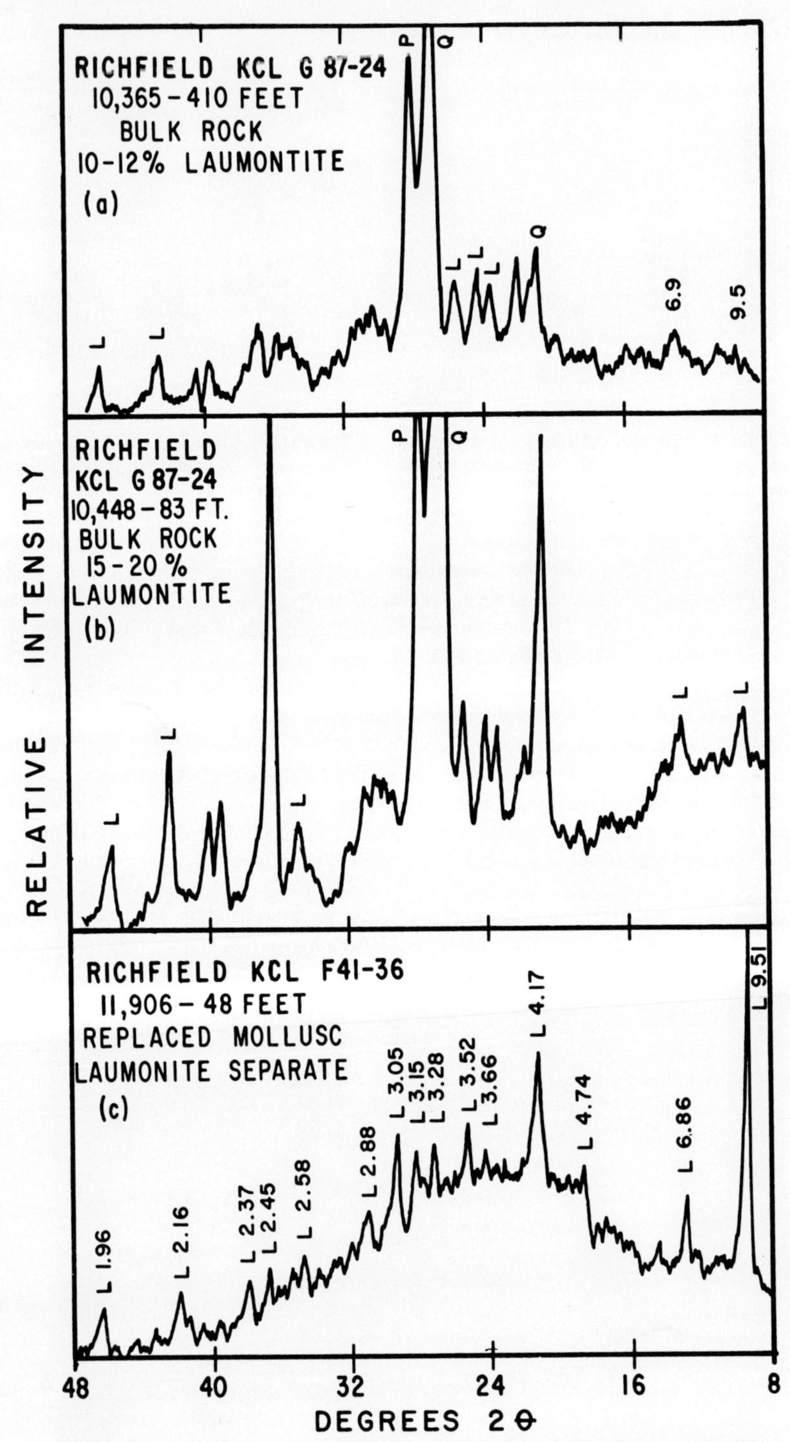


**Figure 12.** X-ray diffraction patterns, Tejon area, California. L, laumontite and leonhardite; P, plagioclase; Q, quartz; *d* spacings in Angstrom units. CuK$\alpha$, $\lambda$ = 1.5418 A.

**Figure 13. Photomicrograph of an unaltered, loosely packed, porous arkosic sandstone. Crossed nicols. Standard CCMO 4-35, 5,238 to 5,263 ft.**

It is recognized that the temperature data from mechanical logs are usually inaccurate. Many factors are responsible for inaccuracy, but the principal one is that temperature equilibrium is not attained during the short time span without mud circulation. Consequently, long-duration tests provide the best bottom-hole temperature information.

Accurate temperature surveys have been used to establish standards by which electrical log data were corrected. Some of these data were obtained from the discovery well of the Swanson River oil field in the Cook Inlet area of south-central Alaska, Standard Oil of California-Richfield Oil Company Swanson River Unit 34-10; the shut-in period was in excess of one hundred days (Fig. 15). On Figure 15 are the recorded temperature readings from this and nearby wells in the Swanson River Field; the correction applied to the electric log data is the difference between the average of the raw temperature data obtained from electric log runs and from the temperature survey. For example, if the recorded temperature is 140°F (60°C) the corrected temperature is 168°F (76°C). The relation established in Cook Inlet, Alaska, has proven to be applicable to other areas where large diameter boreholes (six inches or larger) are normally drilled, such as the Los Angeles and San Joaquin Basins of California.

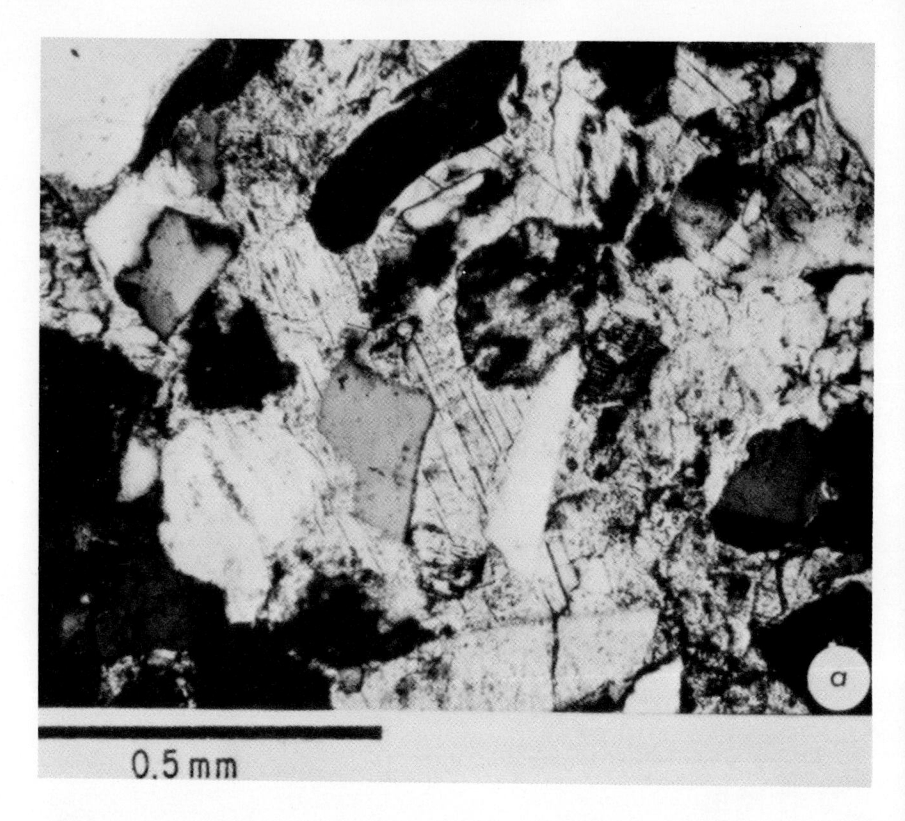


**Figure 14. (a) Arkosic sandstone with laumontite completely filling the pore space and replacing detrital grains; 10 to 15 percent laumontite; 10,183 to 10,226 ft. (b) Laumontite occurring**

In the Tejon area, we used both corrected electric log temperatures and tests of long duration, ranging up to thirty days. Both sources of data agree rather well (Fig. 16). A linear regression analysis yields a geothermal gradient of 1.28°F per 100 ft (2.3°C per 100 m) through the interval studied; the average gradient, extrapolated to a surface temperature of 70°F (21°C), is 1.37°F per 100 ft (2.5°C per 100 m).

The present-day subsurface conditions in the Tejon area can be used to determine the temperature and pressure required for the formation of laumontite, since the petrographic study showed that it is a postcompaction authigenic mineral. Authigenic mineral assemblages are relatively sensitive to pressure and temperature, because zeolites are hydrous mineral phases of low specific gravity (Hay, 1966). As the rocks in the Tejon area are porous and permeable, the subsurface pore fluid pressure is essentially hydrostatic. The pressure in the 9,500- to 10,500-ft interval is therefore about 4,200 to 4,600 psi (290 to 320 bars), or less than half the load pressure.

The average reflectance in the 9,500- to 10,500-ft section is 0.44 to 0.46 percent, which, according to our calibration (Fig. 1), falls within the subbituminous A to high-volitile C bituminous coal rank.

**as a cement and as a replacement of detrital grains; 20 percent laumontite; crossed nicols. Richfield KCL G 76-24, 10,962 to 11,020 ft, North Tejon field.**

---

**Cache Creek Area, California**

**Introduction.** The Great Valley sequence of late Mesozoic age crops out along the west side of the Great Valley of California (Fig. 17). It consists of some 45,000 ft of marine clastic sediments which form an easterly dipping homoclinal succession (Rich and others, 1968; Dickinson and others, 1969). The coeval Franciscan assemblage includes rocks of the zeolite, blueschist, and eclogite facies (Bailey and others, 1964). In the Cache Creek area, Dickinson and others (1969) reported the presence of laumontite as a characteristic alteration product in Lower Cretaceous sandstones of the Great Valley sequence. The estimated maximum burial depth for the top of the exposed section is around 5,000 ft, since beds of Campanian to Maestrichtian age which are present in subsurface to the southeast are presumed to have been deposited in this area as well. In the Cache Creek area, the Great Valley sequence was uplifted in earliest Tertiary time and is unconformably overlain by the lower Eocene Capay shale. Therefore, the duration of maximum burial for the Great Valley sequence was brief, probably less than 5 m.y.

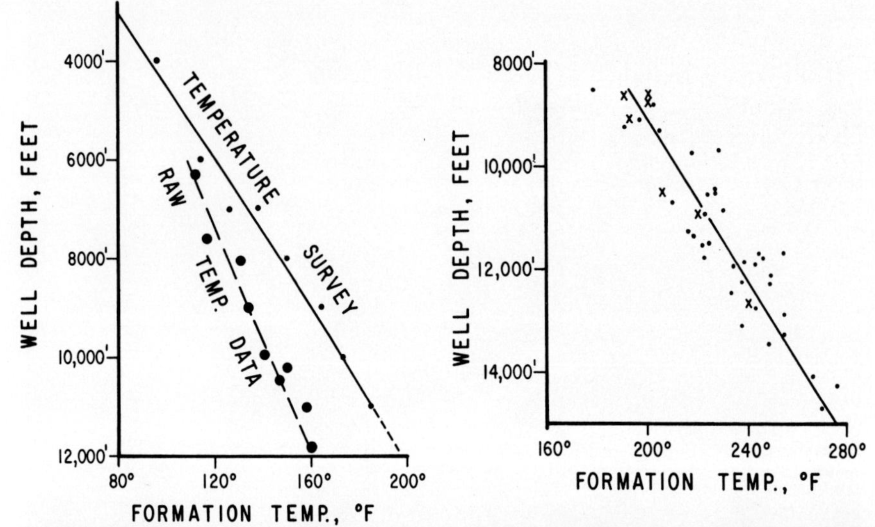


**Figure 15. Formation temperature-depth relation, Swanson River field, Alaska. The raw temperature data were obtained from maximum reading thermometers accompanying electrical log surveys. The temperature survey was run after a shut-in period of 100 days.**

**Figure 16. Tejon area, temperature-depth plot, with linear regression best-fit line. Crosses are from long duration tests; dots are corrected temperatures from electrical logs.**

**Vitrinite Reflectance Study.** Samples for vitrinite reflectance study were taken from three areas centering on Cache Creek (Fig. 17). Shell's Putah Creek diamond corehole section provided a well-measured, excellent series of unweathered samples through most of the Upper Cretaceous beds. The reflectance data from the coreholes line up well with the data obtained from outcrop (Fig. 18). Samples with a high mean reflectance generally exhibit weathering characteristics, and they were not used in making the interpretation.

Figure 19 shows histograms from adjacent samples in the Yolo Formation which, in this locality, contains many thin turbidite sequences. The core from 71 to 72 ft is a pebbly mudstone with large carbonaceous fragments and reworked shale clasts up to 5 cm (2 in.) in length; the core sample from 73 to 77 ft is a shale without any visible reworked components. The histogram from the 71- to 72-ft interval has a wide spread in reflectance and lacks any clear cut mode, while the 73- to 77-ft core has a very strong unimodal distribution. Only a small part of the vitrinite from the shallower depth is considered to be first cycle; most of it comes from clasts. The contrasting results from these closely spaced core samples in the Yolo Formation emphasize the importance of depositional environment and lithology on the amount and character of vitrinite present in a sediment. Except for the sample cited above, which was run expressly to illustrate the problems connected with reworked vitrinite, we excluded all rocks that contained readily recognized reworked material.

The principles outlined above assisted in establishing the preferred interpretation shown on Figure 18. The sample from the Fiske Creek Formation (see Fig. 20) has a multimodal vitrinite distribution, and only the lowest reflectance group is judged to be unaltered and in place; the higher reflectance vitrinite (reflectance in oil, that is, Ro greater than 0.65 percent) does not seem to belong to this reflectance group, and it is similar in reflectance to the altered and reworked

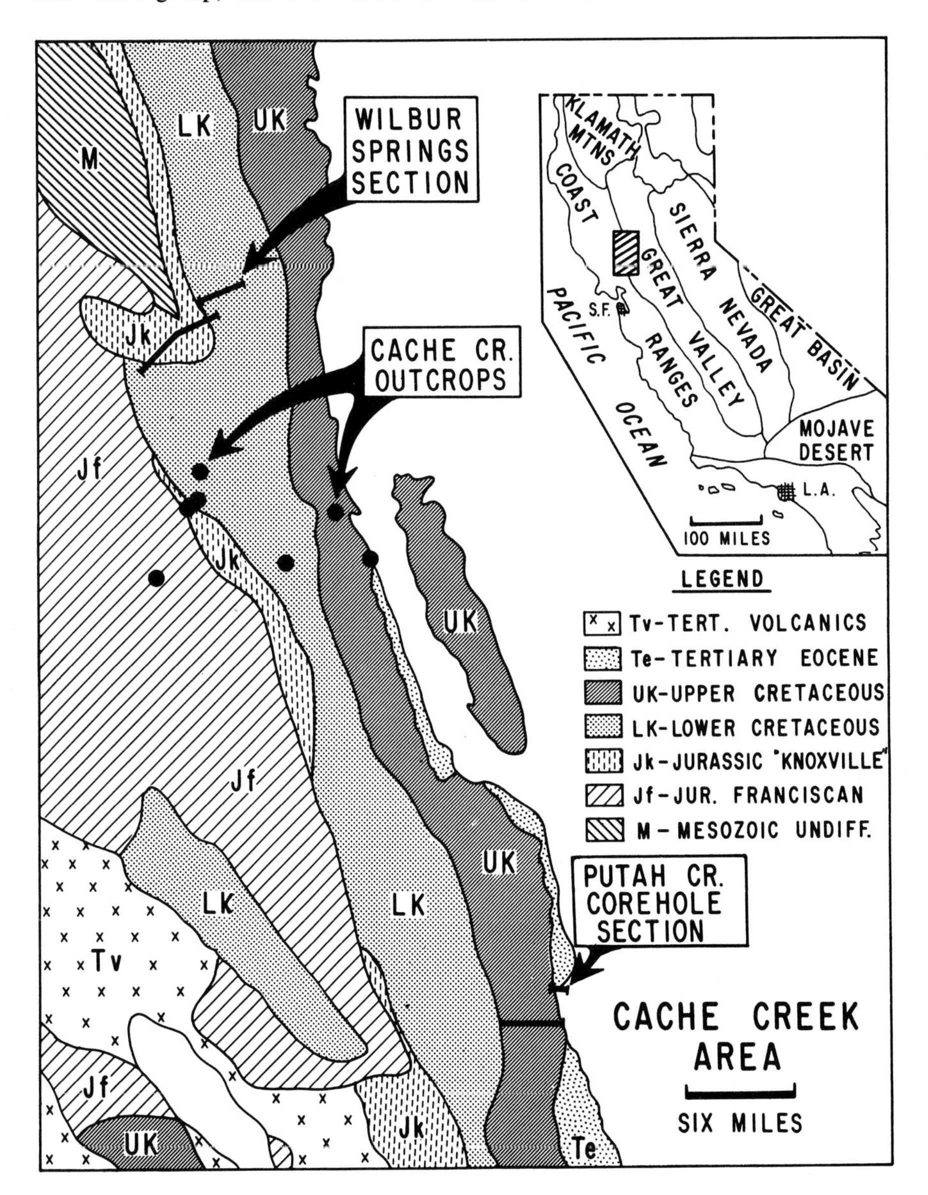


**Figure 17. Generalized geologic map of part of the west side of the Great Valley of California, showing the location of samples described in the text.**

vitrinite present in the Yolo Formation (Fig. 19). In the Crack Canyon Formation, near the base of the Lower Cretaceous section, and with a burial of some 36,000 ft, the mean reflectance is only 0.62 percent (Fig. 21). The strong unimodal vitrinite reflectance distribution is typical of the Crack Canyon and "Knoxville" Formations and is unlike the histograms from the Upper Cretaceous that are sometimes difficult to interpret.

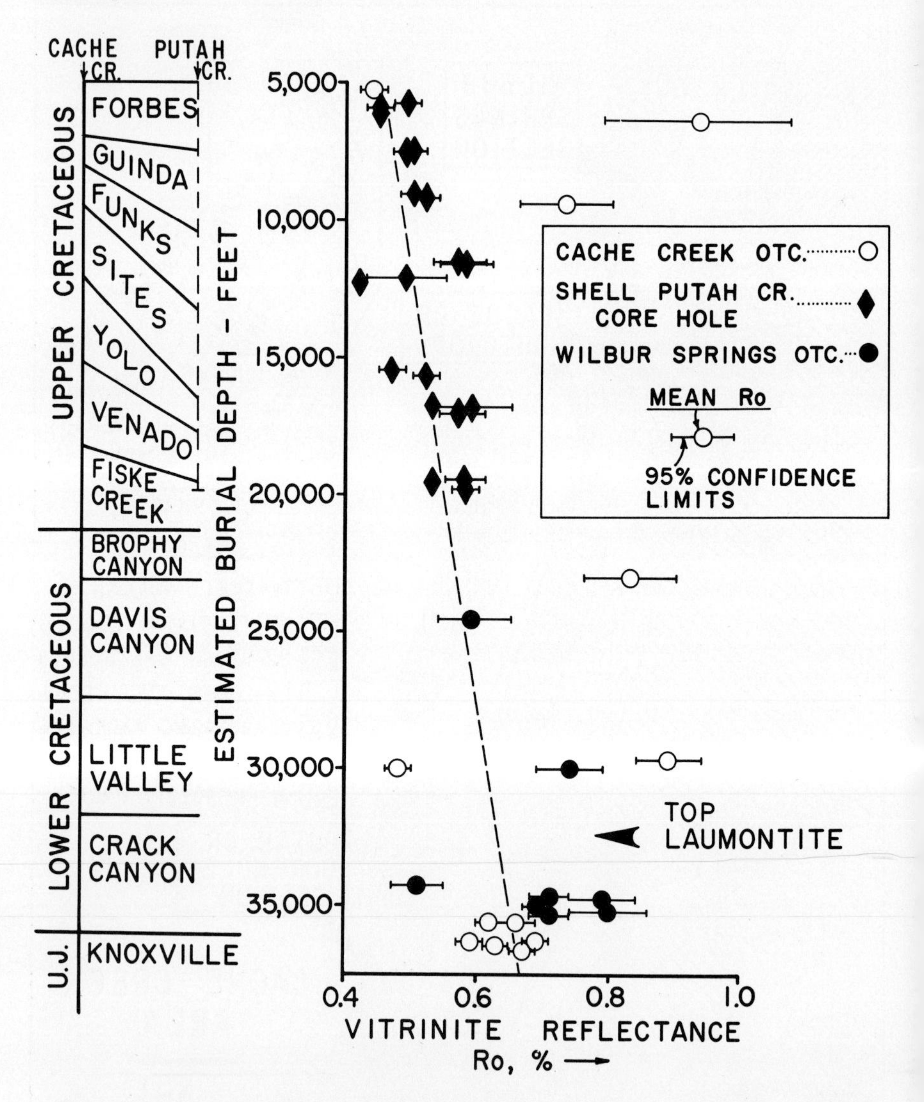


**Figure 18. Reflectance-estimated burial depth plot, Cache Creek area and vicinity, California. Stratigraphic column adapted from Page (1966). At Cache Creek, the Upper Cretaceous layer is thinner than it is at Putah Creek; the Cache Creek thickness is represented on the left side of the stratigraphic column, Putah Creek on the right. For sample locations, refer to Figure 17.**

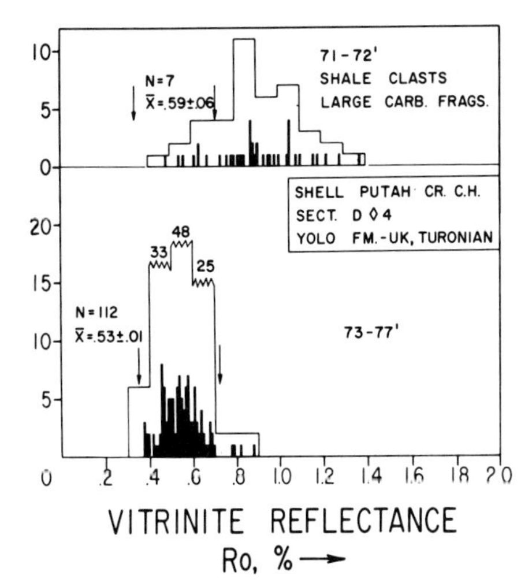


**Figure 19. Vitrinite reflectance histograms, Yolo Formation, Putah Creek core hole section. Estimated burial depth, 16,500 ft.**

**Figure 20. Vitrinite reflectance histogram, Fiske Creek Formation, Putah Creek core hole section. Estimated burial depth 19,000 ft.**

**Development of Authigenic Minerals.** Based on x-ray analysis of bulk-rock samples, Dickinson and others (1969) found that albitized sandstones from the Crack Canyon Formation with an estimated burial depth of 32,500 ft showed clear evidence for the presence of laumontite. In addition, they found samples higher in the section that also yielded peaks suggestive of laumontite and inferred the occurrence of laumontite in sandstones with burial depths greater than 17,500 ft.

From a study of thin sections from the Great Valley sequence, we were able to verify the presence of laumontite in the Crack Canyon Formation, and x-ray diffraction analysis of mineral separates confirmed the mineral identification. However, we did not find laumontite in thin section at stratigraphic levels higher than a burial depth of 32,500 ft. In the Tejon area, it was shown that laumontite peaks do not show up well in x-ray diffraction patterns of bulk rock samples even where laumontite makes up 10 to 12 percent of the sandstone (Fig. 12). The recognition of laumontite can be made reliably from thin section in concentrations considerably smaller than 10 percent, and the x-ray analysis on mineral separates

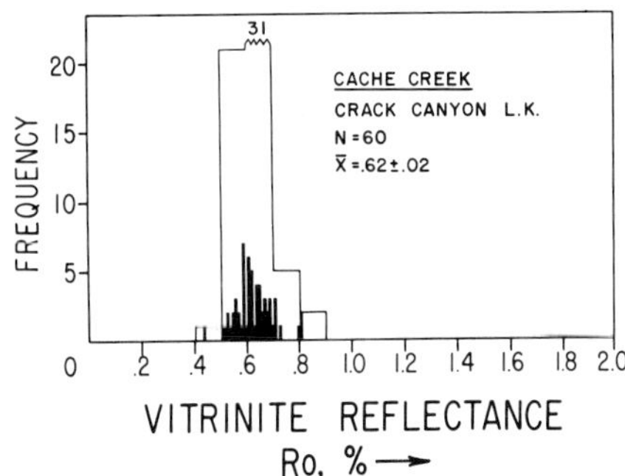


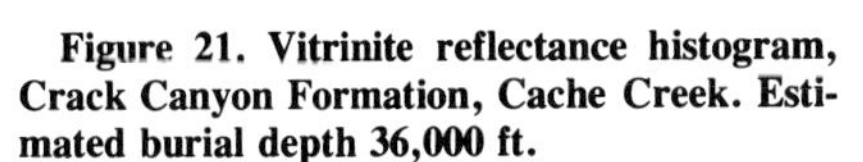

**Figure 21. Vitrinite reflectance histogram, Crack Canyon Formation, Cache Creek. Estimated burial depth 36,000 ft.**

can then be used for positive identification. It is concluded that, by itself, x-ray diffraction analysis of bulk rock samples should not be relied upon for the determination of laumontite.

Analogous to the Tejon area, laumontite is a postcompaction authigenic mineral in the Great Valley sequence, but it makes up at most 5 to 6 percent of the rock compared with 15 to 20 percent in the Eocene rocks at Tejon. Higher permeability and a continuous flow of ions into the system favor a more complete reaction (Hay, 1966); this could account for the higher percentage of laumontite in the more permeable rocks at Tejon.

The fluid pressures prevailing in the Great Valley sequence at maximum burial depth were reconstructed by making two assumptions: (1) the fluid pressure in the more porous and permeable section would have been essentially hydrostatic, or approximately 0.44 psi per ft, (0.10 bar per m), and (2) the pressures in the impermeable section would have approached load pressure, or about 1.0 psi per ft (0.23 bar per m). Porous and permeable sandstones are present in the uppermost 8,000 ft of the Great Valley sequence at Cache Creek, down to the base of the Guinda Formation; through the remainder of the section, the highest permeability measured was 0.5 millidarcy. Based on these assumptions, at maximum burial, the pressures at the top of the laumontite-bearing zone were around 27,000 psi (1,860 bars).

The average reflectivity of vitrinite is 0.63 percent at the top of laumontite mineralization (Fig. 18). This is equivalent to a high-volatile B bituminous coal rank.

Although the Great Valley sequence is older than the sediments present in the Tejon area, the time span involved in the maximum burial phase is comparable for both areas, probably less than 5 m.y. At Cache Creek, the reflectance at the first occurrence of laumontite is 0.63 percent; in the Tejon area, the temperature at this reflectance level is 265°F (130°C). By using the chart devised by Karweil (1956) in the Tejon area (Fig. 6), we find that the time factor for a coal rank equivalent to 0.63 percent reflectance and 130°C (265°F) is 15 m.y.; these beds are of upper Eocene age, approximately 50 m.y. old. If we had used 50 m.y. and the measured reflectance to obtain the time factor, the calculated temperature would be 90°C (194°F), 40°C (72°F) less than the measured bottom-hole temperature for this coal rank. We conclude that the time-temperature relations of Karweil based on reaction kinetics more closely fit our model of a short time of maximum heating at Tejon, which lends validity to the approach that we have taken for the Cache Creek area.

If we accept the concept of a brief maximum heating phase at Cache Creek, then the data from the Tejon area can be applied directly, and may be used as a geological thermometer for the Great Valley sequence.

At Cache Creek, the depth of burial at the top of laumontite mineralization is estimated at 32,500 ft. If we use a temperature of 265°F (130°C), the calculated paleogeothermal gradient is 0.62°F per 100 ft (1.1°C per 100 m), assuming a surface temperature of 70°F (21°C). If we had used a time factor of 75 m.y. (the approximate age difference between the top of the restored Great Valley sequence and the laumontite-bearing interval), the calculated paleotemperature would be around 170°F (77°C), which yields an even lower paleogeothermal gradient of 0.31°F per 100 ft (0.6°C per 100 m).

The Great Valley sequence was deposited as a thick turbidite wedge on volcanic crust (Bailey and others, 1970) adjacent to the oceanic trench; the calculated gradient is consistent with a model of low heat flow in this tectonic setting (Takeuchi and Uyeda, 1965; Dewey and Bird, 1970). The very low calculated paleogeothermal gradient is in marked contrast with the gradient of 1.5° to 2.0°F per 100 ft (2.7° to 3.6°C per 100 m) measured in wells in the Sacramento Valley east of the Cache Creek area, which penetrate rocks of the same age deposited on crystalline and metamorphic rocks in a normal heat-flow regime.

## FACTORS GOVERNING THE FORMATION OF LAUMONTITE

From this study, we have found that the conditions controlling the formation of laumontite were quite different in the two study areas. At the top of the laumontite-bearing interval in the Tejon area, the measured temperature is 205°F (96°C), and the pore fluid pressure is 4,200 psi (290 bars); in the Cache Creek area we calculated a temperature of 265°F (130°C) and a pressure of 27,000 psi (1,860 bars). As outlined by Hay (1966), zeolitic reactions can proceed at much lower temperatures than in laboratory experiments if fluid pressure is roughly one-third of load pressure, a situation which prevails in the Tejon area. High permeability and a continuous flow of ions favors a more complete reaction; this could account for the higher percentage of laumontite found at Tejon compared with the percentage in the impermeable Great Valley sequence. Burial metamorphic mineral assemblages are dependent on the interrelation between temperature, pressure, and fluid flow; the reaction conditions are also affected by the mineralogical composition, the pore fluid chemistry, and the partial pressure of $CO_2$. In both areas studied, there was an abundance of plagioclase available for alteration; no attempt was made to evaluate the factors other than temperature and fluid pressure.

In an extensive survey of the relation of coal rank to burial metamorphic mineral facies, Kisch (1969) found that laumontite occurred in abundance in coal-bearing rocks with a volatile matter content of less than 40 percent; these are coals of high-volatile A bituminous or higher rank. We learned that in non-coal-bearing rocks, laumontite alteration extends to an equivalent rank of subbituminous A to high-volatile C bituminous coal.

## CONCLUSIONS

Our studies have shown that vitrinite reflectance, as calibrated with ASTM coal rank, can be used successfully in non-coal-bearing sequences as a ranking parameter of burial metamorphic history. Interpretation of the reflectance data depends on sound petrographic analysis and also on a knowledge of lithology and depositional environment to help sort out the disturbing influences of oxidation, bioturbation, and the reworking of second-cycle materials. Because there is an inherent scatter in the mean reflectance data, a sequence of samples is needed to calculate the average reflectivity.

Other workers have demonstrated that the coalification process is largely controlled by temperature and to a lesser degree by time; the effect of pressure is minimal. Electrical log temperature data, as corrected by comparison with accurate subsurface temperature profiles, can be used to establish the present-day temperature regime and permit measurement of paleothermal conditions.

Two areas in California provide a comparison of the reconstruction of burial history utilizing authigenic minerals and vitrinite reflectance. In the Tejon area, laumontite is a postcompaction authigenic mineral that occurs largely as a replacement of plagioclase feldspars and as a cement in arkosic sandstones of lower Miocene to Eocene age. Laumontite is first noted at around 9,500 ft, and below 10,500 ft it makes up as much as 15 to 20 percent of the rock. X-ray diffraction patterns of bulk rock samples did not provide positive identification even when 10 to 12 percent laumontite was present. The average vitrinite reflectance in the 9,500- to 10,500-ft section is 0.44 to 0.46 percent, which is equivalent to subbituminous A high-volatile C bituminous coal rank. Within this depth range, the subsurface temperature is 205° to 218°F (96° to 104°C), and as the section is permeable, the pressure is essentially hydrostatic, or 4,200 to 4,600 psi (290 to 320 bars).

In the Great Valley sequence of late Mesozoic age near Cache Creek, we verified the presence of postcompaction laumontite in the Lower Cretaceous Crack Canyon Formation, as reported by Dickinson and others (1969), but we did not find any laumontite at higher stratigraphic levels as suggested by these authors. Rather than relying on x-ray analysis of bulk rock samples, we use petrographic methods to localize zeolite-bearing intervals and then confirm identifications with x-ray study of mineral separates. Laumontite first occurs at about 32,500 ft below the top of the restored Great Valley sequence. As the section is dominantly impermeable, we assumed that load pressure prevailed through most of the sequence at maximum burial, and the calculated paleopressure at 32,500 ft is 27,000 psi (1,860 bars). The average vitrinite reflectance at the top of the laumontite-bearing interval is 0.63 percent, which is within high-volatile B bituminous coal rank. As the maximum burial phase at both Tejon and Cache Creek is probably less than 5 m.y., we used the Tejon area temperature data as a geological thermometer for the Great Valley sequence. The reflectance level found at Cache Creek at the top of the laumontite-bearing zone corresponds to a subsurface temperature of 265°F (130°C) at Tejon. The calculated paleogeothermal gradient for Cache Creek is 0.62°F per 100 ft (1.1°C per 100 m), which contrasts with a gradient of 1.28°F per 100 ft (2.3°C per 100 m) at Tejon. The low temperature gradient for the Great Valley sequence is consistent with a model of low heat flow for a thick turbidite wedge located adjacent to an oceanic trench.

Because the formation of zeolites is governed by a complex interaction of physical and chemical factors, we judge that vitrinite reflectance, which is largely temperature dependent and not retrogressive in behavior nor sensitive to intrastratal solutions, is the more promising tool for the measurement of thermal history.

## ACKNOWLEDGMENTS

Thanks are extended to the Shell Oil Company for permission to publish this paper. Peter Van de Kamp conducted the x-ray analyses, examined some of the thin sections, and collected a number of the outcrop samples. We are especially indebted to Ervin Cain, who performed the vitrinite reflectance analyses. The figures were drafted by Holger Bystrom and Kent Cardon. Standard Oil Company of California and the Atlantic Richfield Company granted permission for us to publish the Swanson River field temperature survey. The manuscript was critically read by Kaspar Arbenz and Darrel Cowan.

## REFERENCES CITED

American Society for Testing and Materials, 1970, 1970 Book of ASTM standards: Pt. 19, Gaseous fuels, coal and coke, Philadelphia, Pennsylvania, 502 p.

Bailey, E. H., Irwin, W. P., and Jones, D. L., 1964, Franciscan and related rocks, and their significance in the geology of western California: California Div. Mines and Geology Bull. 183, 177 p.

Bailey, E. H., Blake, M. C., Jr., and Jones, D. L., 1970, On-land Mesozoic oceanic crust in California: U.S. Geol. Survey Prof. Paper 700-C, p. C70-C81.

Benedict, L. G., and Berry, W. F., 1966, Further applications of coal petrography, *in* Given, Peter, Chm., Coal science: Washington, D.C., Advances in Chemistry Ser. No. 55, Am. Chem. Soc., p. 577-601.

Bostick, N. H., 1970, Thermal alteration of clastic organic particles (phytoclasts) as an indicator of contact and burial metamorphism in sedimentary rocks [Ph.D. thesis]: Stanford, Calif., Stanford Univ., 220 p.

Dean, R. B., and Dixon, W. J., 1951, Simplified statistics for small numbers of observations: Anal. Chemistry, v. 23, p. 636-638.

de Vries, H.A.W., Habets, P. J., Bokhoven, C., 1968, Das Reflexionsvermögen von Steinkohle II, Die Reflexionsanisotropie: Brennstoff-Chemie, v. 49, p. 47-52.

Dewey, J. F., and Bird, J. M., 1970, Mountain belts and the new global tectonics: Jour. Geophys. Research, v. 75, p. 2625-2647.

Dickinson, W. R., Ojakangas, R. W., and Stewart, R. L., 1969, Burial metamorphism of the late Mesozoic Great Valley sequence, Cache Creek, California: Geol. Soc. America Bull., v. 80, p. 519-525.

Gray, R. J., and Shapiro, N., 1966, Petrographic composition and coking characteristics of Sunnyside coal from Utah, *in* Central Utah coals: Utah Geol. and Mineralog. Survey Bull. 80, p. 55-79.

Hacquebard, P. A., and Donaldson, J. R., 1970, Coal metamorphism and hydrocarbon potential in the upper Paleozoic of the Atlantic provinces, Canada: Canadian Jour. Earth Sci., v. 7, p. 1139-1163.

Hay, R. L., 1966, Zeolites and zeolitic reactions in sedimentary rocks: Geol. Soc. America Spec. Paper 85, 130 p.

Huck, G., and Patteisky, K., 1964, Inkohlungsreaktionen unter Druck: Fortschr. Geologie Rheinland u. Westfalen, v. 12, p. 551-558.

Kaley, M. E., and Hanson, R. F., 1955, Laumontite and leonhardite cement in Miocene sandstone from a well in San Joaquin Valley, California: Am. Mineralogist, v. 40, p. 923-925.

Karweil, J., 1956, Die Metamorphose der Kohlen vom Standpunkt der physikalischen Chemie: Deutsch. Geol. Gesell. Zeitschr., v. 107, p. 132-139.

Kisch, H. J., 1969, Coal rank and burial-metamorphic mineral facies, *in* Schenk, P. A., and Havenaar, I., eds, Advances in organic geochemistry, 1968: Oxford, England, Pergamon Press, p. 407-425.

Page, B. M., 1966, Geology of the Coast Ranges of California: California Div. Mines and Geology Bull. 190, p. 255-276.

Park, W. H., 1961, North Tejon oil field: California Oil Fields—Summ. operations, v. 47, no. 2, p. 13-22.

Rich, E. I., Ojakangas, R. W., Dickinson, W. R., and Swe, W., 1968, Sandstone petrology of Great Valley sequence, Sacramento Valley, California: Geol. Soc. America Spec. Paper 121, p. 550.

Takeuchi, H., and Uyeda, S., 1965, A possibility of present-day metamorphism: Tectonophysics, v. 2, p. 59.

Teichmüller, Marlies, and Teichmüller, Rolf, 1966, Geological causes of coalification, *in* Given, Peter, Chm., Coal science: Washington, D.C., Advances in Chemistry Ser. 55, Am. Chem. Soc., p. 133-155.

——1968, Geological aspects of coal metamorphism, *in* Murchison, D., and Westoll, T. S., eds., Coal and coal-bearing strata: Edinburgh, Oliver & Boyd, p. 233-267.

SYMPOSIUM HELD AT G.S.A. ANNUAL MEETING IN MILWAUKEE, NOVEMBER 1971
MANUSCRIPT RECEIVED BY THE SOCIETY MARCH 9, 1973

AUTHOR'S PRESENT ADDRESS: CASTAÑO—SHELL OIL COMPANY, P.O. BOX 481, HOUSTON, TEXAS 77001; SPARKS—GEO-LOGIC, 436 W. COLORADO BOULEVARD, GLENDALE, CALIFORNIA 91204

Printed in U.S.A.

Geological Society of America
Special Paper 153


# Coalification Patterns of Pennsylvanian Coal Basins of the Eastern United States

Heinz H. Damberger

*Illinois State Geological Survey*
*Urbana, Illinois 61801*

## ABSTRACT

Coalification patterns of Pennsylvanian coal basins in the eastern United States reflect (1) depth of burial during later Pennsylvanian and Permian times, when the main coalification took place; and (2) regional thermal disturbances from Permian to recent times.

In most of the eastern United States, rank was determined during the main phase of coalification in Pennsylvanian and Permian times and thus reflects former greatest depths of burial. The general decrease in rank toward the Canadian Shield corresponds well to the general northward thinning of the mantle of sediments over the basement. In the coalification map of a reference seam, basins (for example, Illinois Basin) show up as northward extensions of high-rank areas, and positive structures (for example, Cincinnati Arch, Ozark Uplift) appear as southward protrusions of low-rank areas. The sedimentary column over the reference seam, or its stratigraphic equivalents, was thicker within the basins than over the positive structures.

If data on rank are plotted on maps for coal seams lying at or near the surface regardless of their stratigraphic age, the close relation to the large structural units is much less pronounced, and the geologic interpretation of such coalification maps is more difficult.

Superimposed upon this broad-scale and rather simple coalification pattern that reflects burial depth are several areas of anomalously high rank that are interpreted as being the result of unusually high heat flow sometime after the main coalification had terminated, probably in connection with deep-seated plutonic activity. These comprise the Rhode Island Meta-anthracite region, the Pennsylvania Anthracite region extending westward into the area of low- and medium-volatile bituminous coals of Pennsylvania and northern West Virginia, the eastern portion of the Arkoma Basin in Arkansas, and possibly also the southern portion of the Illinois Basin and the area of high-rank bituminous coals in southern West Virginia. In the last two regions, the high rank may be explainable solely by greater former depths of burial. The coalification patterns of all these areas are considered anomalous compared to the patterns of adjacent areas and of other coal basins. Supporting evidence for this interpretation of the coalification pattern is furnished by the presence in the same areas of other manifestations of large-scale regional heating and deep-seated plutonic activity.

## INTRODUCTION

Regional and local coalification studies yield valuable information on the relative age of burial, folding, faulting, and paleogeothermic events. They are also an excellent tool for the areal evaluation and prediction of such important coal properties as heating value, coking and caking ability, and moisture and fixed-carbon contents.

The coalification pattern that is now observed in sediments may be the result of one single geologic event (for example, burial) or, more likely for coals as old as Pennsylvanian, a sequence of geological events, including burial, folding, faulting, igneous intrusion, and renewed burial. Very often, the detailed study of regional and local patterns of coal metamorphism has helped unravel complex geologic histories.

Organic accumulations such as peat are transformed into coal during burial under younger sediments. The highest rank of coalification will normally be reached during the period of greatest depth of burial. The coalification pattern prior to any appreciable folding and faulting is termed predeformational or preorogenic. Surprisingly enough, this stage has been preserved in many areas, even where intense folding and faulting occurred later. Most of the Ruhr Basin of western Germany is a classic example of the preservation of the preorogenic coalification pattern during intensive folding, thrusting, and normal and shear faulting. The lowest rank in the Sonnenschein coal seam is found along the southern margin of the Ruhr district, where folding and thrusting are most prominent, but burial of this seam was minimal (Fig. 1).

The coalification process may not have come to an end before folding and thrusting took place. The Merlenbach anticline and the adjacent Marienau syncline of the Saar-Lorraine coal district are good examples of continued coalification during folding (Fig. 2). In that area, thrusting also started while coalification was still in progress (Fig. 3). Normal faulting, however, clearly postdates coalification—both marker beds and isoranks are displaced by the same amount. However, while this is so in many coal fields, it is not always the case. Skipsey (1959) reported that in the Stirlingshire coal field of Scotland where normal faulting occurred while coalification was still going on, rank within a seam is always higher on the downthrown side of the major normal faults, which have vertical displacements of several hundred feet.

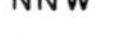

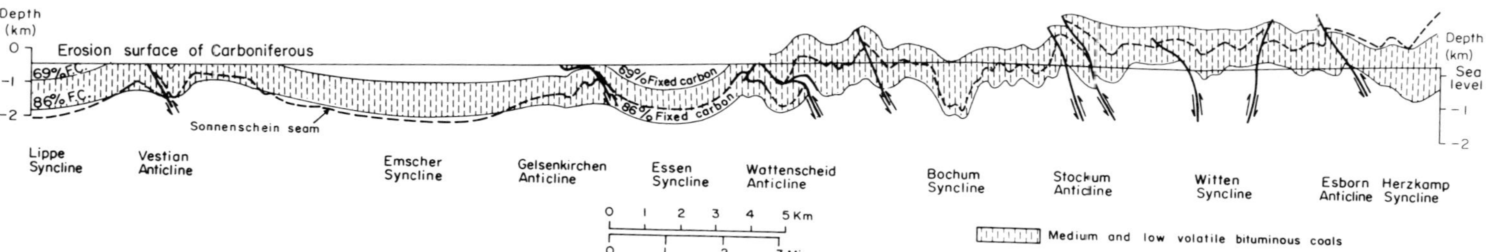


**Figure 1.** Cross section through the Ruhr Basin, West Germany, showing the Sonnenschein coal seam and the low- and medium-volatile bituminous range of coalification (hachured), to depict interrelation between folding and faulting of beds and isoranks (after Patteisky and others, 1962).

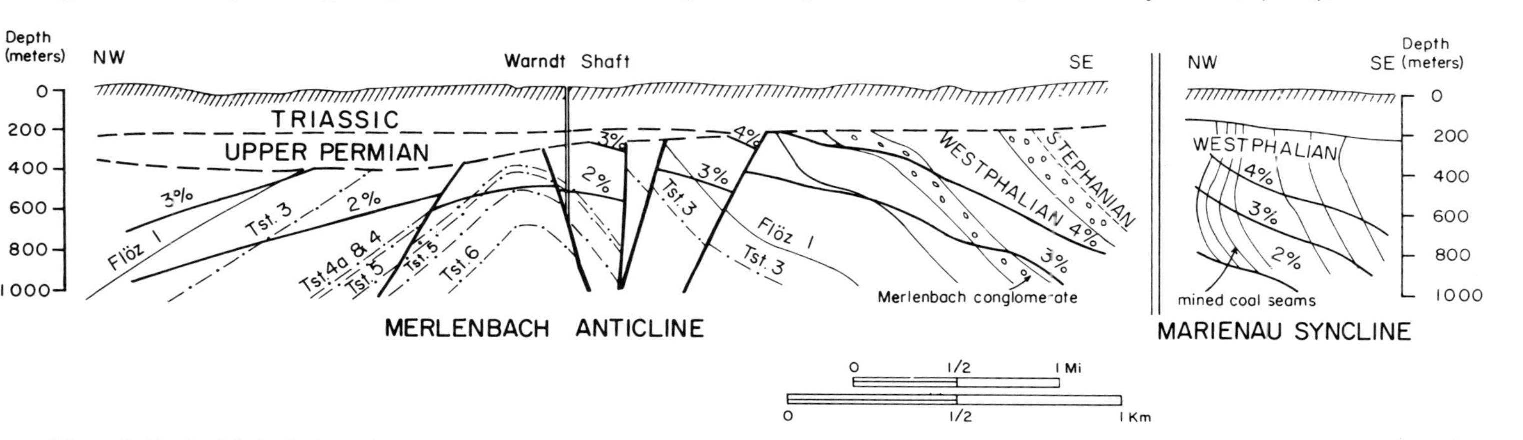


**Figure 2.** In the Merlenbach anticline, Saar coal district, West Germany, and in the northwestern limb of the adjacent Marienau syncline, Lorraine, France, the isoranks exhibit less folding than the strata, indicating that the folding and maturing of the coal were simultaneous. Rank is depicted by averaged hygroscopic moisture content, ash-free basis, which proved to be a good parameter of rank in these high-volatile bituminous coals (from Damberger, 1966b). Tst. = Tonstein.

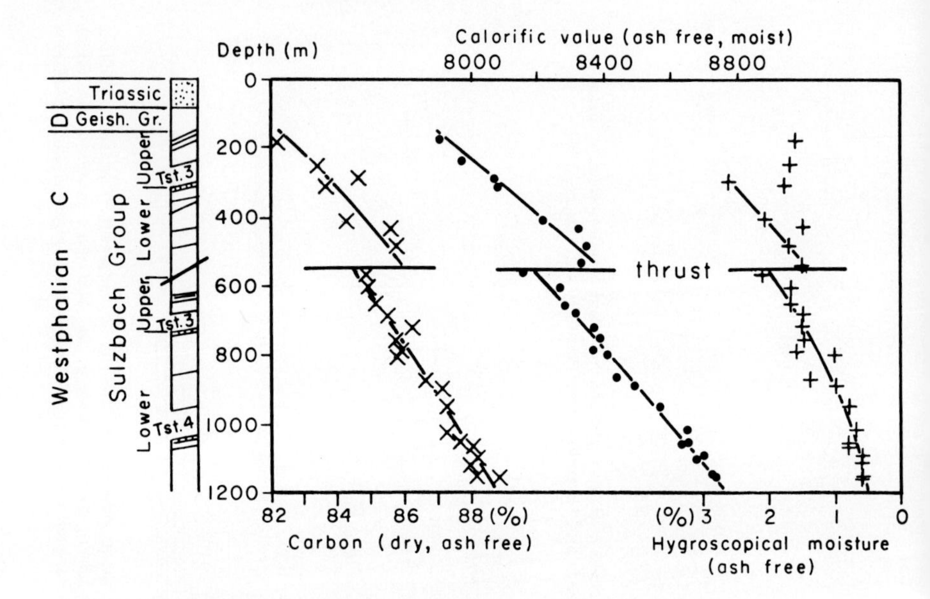


**Figure 3. Borehole Kughütte 2, Saar coal district. A major thrust that repeats about 450 m of the stratigraphic section (interval from Tonstein 3 to Tonstein 3) leads to the repetition, after only about 180 m, of coal of the same rank (for example, 8,300 kcal per kg at 490 and 670 m), thus indicating a partial overlap in time of the coalification process and thrusting (from Damberger and others, 1964; interpretation after Damberger, 1966b). Tst. = Tonstein.**

The coalification map of the Sonnenschein seam of the Rhine-Ruhr coal district depicts a general increase in rank toward Krefeld, west of the Rhine River (Fig. 4). The increase cannot be attributed to greater former depth of burial, because Carboniferous formations in this area tend to thin to the west. Patteisky and others (1962) suggested that the mineralization, igneous intrusions, and high rank of the coals, all present in this area, are related to the same cause—an extensive magmatic upwelling at depth in this western portion of the Rhine-Ruhr coal district. Teichmüller and Teichmüller (1966a) called this action telemagmatic coalification. An even more conspicuous geothermal impact on the coalification pattern is associated with a large, basic intrusive body near Erkelenz in the westernmost part of the district (Fig. 4). Another well-studied example of telemagmatic coalification above a deep-seated igneous intrusion is that of the Brahmsche Massif of northwestern Germany (Fig. 5).

## REGIONAL COALIFICATION PATTERNS OF PENNSYLVANIAN COAL BASINS OF THE UNITED STATES

Published coalification maps of the Pennsylvanian coal basins of the United States generally have been based on data from near-surface coals, regardless of their stratigraphic positions (White, 1915; Fuller, 1919, 1920; Semmes, 1920, 1929; Eby, 1923; Moulton, 1925; Croneis, 1927; Campbell, 1930; Miser, 1934; Hendricks,

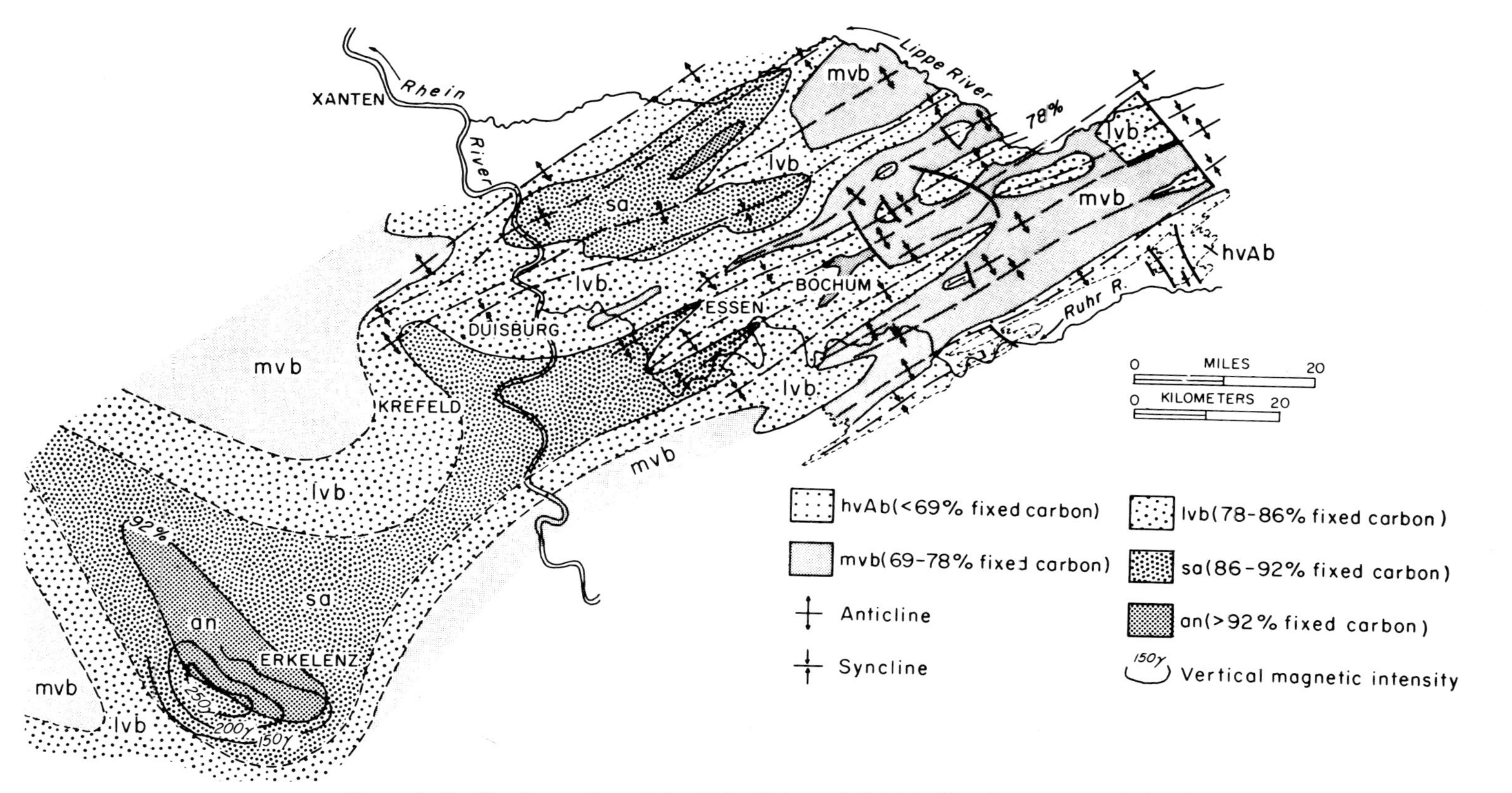


**Figure 4.** Coalification pattern of the Rhine-Ruhr coal district, West Germany, as depicted by isoranks on the Sonnenschein coal seam (after Patteisky and others, 1962; Teichmüller and Teichmüller, 1968b).

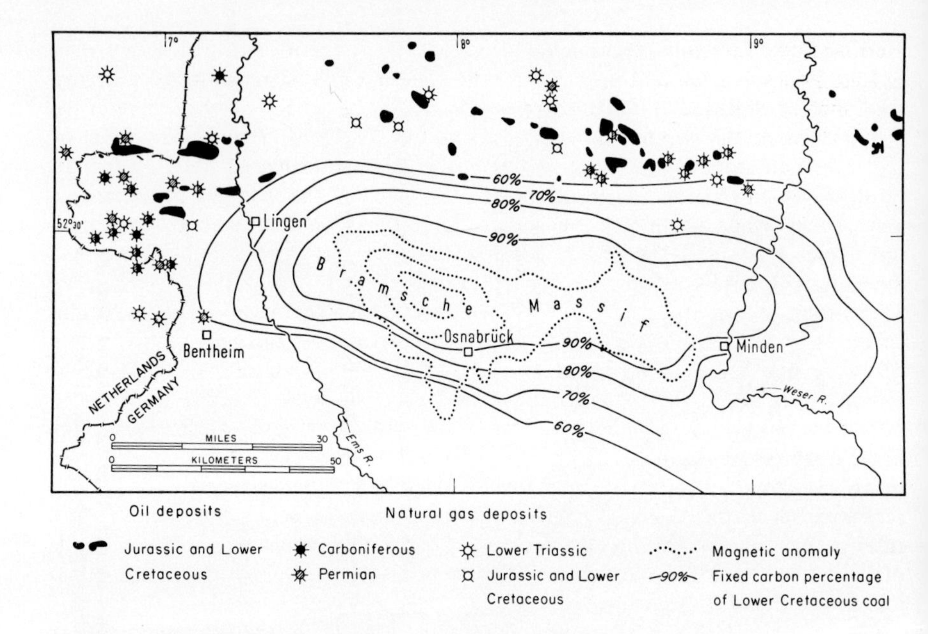


**Figure 5. The Brahmsche Massif of northwestern Germany, a deep-seated basic igneous intrusion, is well delineated by a magnetic anomaly as well as by anomalously high-rank coals of both Cretaceous and Carboniferous ages. The isoranks are on the Lower Cretaceous coals, but the Carboniferous coals near Osnabrück are also high-rank bituminous and anthracitic coals. Oil fields are restricted to carbon ratios below 70 to 60. The isoranks are believed to represent approximately equal temperatures during the cooling period of the Brahmsche pluton (from Teichmüller and Teichmüller, 1968a).**

1935; Trumbull, 1960). The geologic interpretation of such coalification maps is more complicated than the interpretation of maps for single coals regardless of depth, and comparison with coalification patterns in other coal basins is difficult. An attempt was therefore made to relate all coalification data of the Interior and Eastern Coal Provinces of the United States to approximately the same coal seam (Figs. 6 and 7). The procedure used is described in the Appendix.

The Herrin (No. 6) Coal Member of the Carbondale Formation in Illinois was chosen as the reference seam because of its wide distribution in the Illinois Basin and the wide distribution of its presumed correlative, the Middle Kittanning Coal Member of the Allegheny Group, in the Appalachian Basin (Moore and others, 1944). Although this choice is generally appropriate, it is unfavorable for the Arkoma Basin and the southern Appalachian Basin, where the outcropping and mined coals are much lower in the stratigraphic sequence.

## Isoranks as Indication of Former Depth of Burial

As a first approximation, the isoranks (lines of equal rank) in Figure 6 can be considered to delineate areas of comparable former depths of burial of the

Herrin-Middle Kittanning coal seams during the main period of coalification, probably in Late Pennsylvanian and in Permian time. The over-all decrease in rank toward the Canadian Shield to the north corresponds well to the general northward decrease in thickness of the sedimentary cover above the basement. The deeper burial of the reference seam in the large epicontinental basins compared to the adjacent positive structures produces wide northward swings of the isoranks in the basins and corresponding southward swings of the isoranks on the positive structures; the Illinois Basin with the Ozark Dome to the west and the Nashville Dome and the Cincinnati Arch to the east is a good example (Fig. 6).

The rapid eastward increase in rank of the reference coal seam in Pennsylvania seems to indicate a corresponding rapid increase of the thickness of sediments that were once laid down on top of it—the Conemaugh, Monongahela, and Dunkard Groups, their stratigraphic equivalents, and possibly some younger strata. The difference in rank, that is, high-volatile A and medium-volatile bituminous coals in the Appalachian Basin and anthracites in eastern Pennsylvania, should correspond to an equivalent difference in the former thickness of sediments on top of the reference seam. If a "normal" increase in fixed carbon content with depth (Hilt's rule) is adopted, the *additional* thickness of sediments (equal to *additional* depth of burial) in the anthracite areas of Pennsylvania should amount to some 3,000 to 5,000 ft (Fig. 14).

The average thickness preserved above the reference seam is about 2,500 ft in the bituminous coal fields and 2,500 to 3,200 ft in the anthracite fields (Edmunds, 1971; Wood and others, 1969). Because of the poor preservation of plant fossils in the anthracite fields, the correlation to the type sections in the bituminous coal fields is still rather tentative, especially for the upper portion of the Llewellyn Formation of the anthracite fields. It is possible that more of the sedimentary column was eroded in the anthracite fields than in the bituminous coal fields, possibly even as much as the 3,000 to 5,000 ft, the amount one might expect from the difference in the degree of coalification between the two areas. Wood and others (1969), in a discussion of the coalification pattern of the west-central part of the Southern Anthracite Field, concluded that most of the rank changes in the area can be attributed to differences in the former depth of burial. However, close inspection of the data reveals some rather unusual coalification patterns. In cross sections, rank is highest toward the centers of the two troughs of the Minersville synclinorium, which means that the younger coals within the synclinorium are higher in rank than the older coals on its flanks; it also indicates that the isoranks and coal beds dip in opposite directions (outward from the synclinorium). Furthermore, the folds of the area plunge northeastward. Successively younger coals were therefore used for the coalification map (Wood and others, 1969, Fig. 42, p. 136). Yet, these younger coals are higher in rank than coals 1,000 ft and more lower in the stratigraphic column at a location some 10 to 20 mi farther southwest. Such a relation between the tectonic structure and the regional coalification pattern is rather typical for large-scale post-deformational (telemagmatic?) heating rather than for a pre- or syn-deformational burial-related coalification pattern.

In the bituminous coal area of the Appalachian Region, the most conspicuous feature is the relatively abrupt northward increase in rank of the reference seam in north-central West Virginia, in Maryland, and in southern Pennsylvania, *parallel* to the structural trend (Fig. 6). Normally, rank variations within one seam are

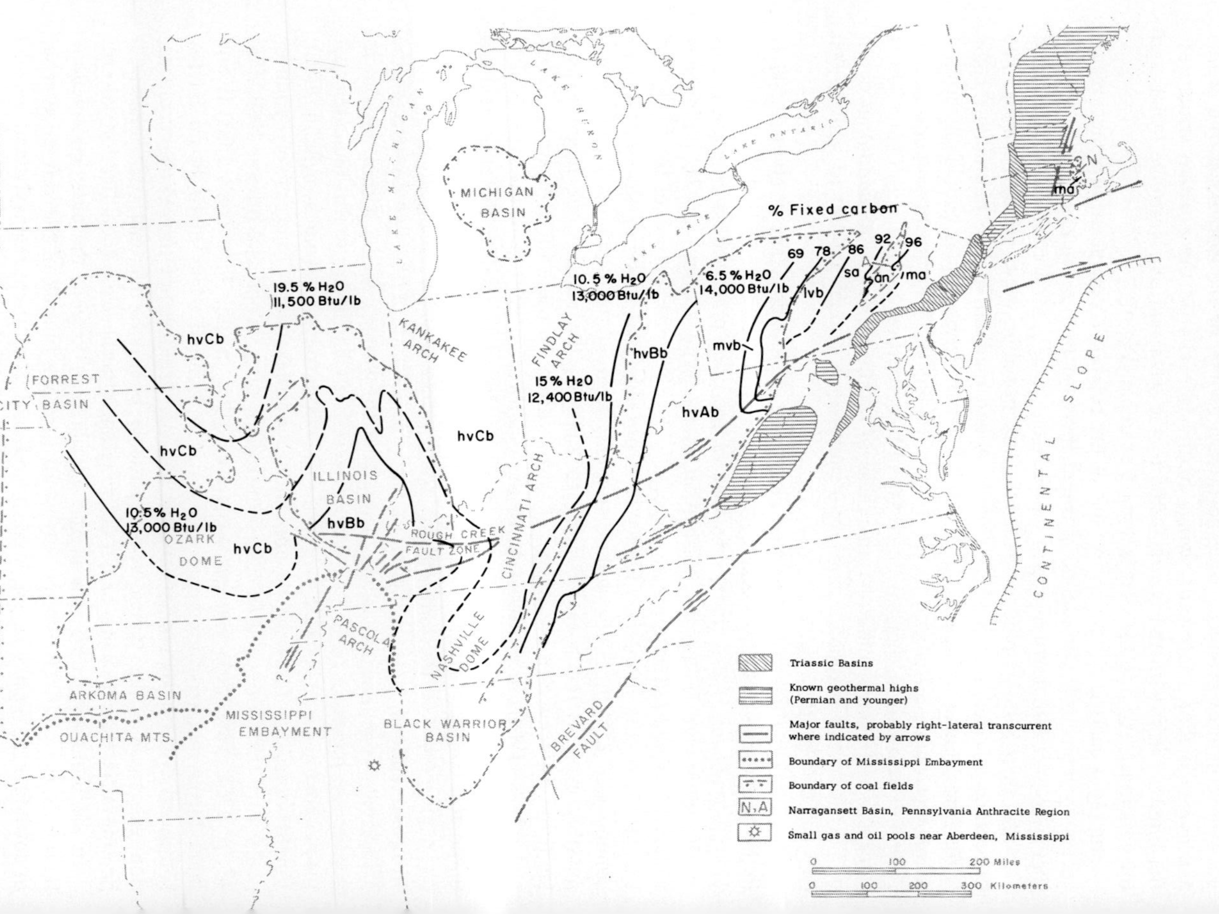


**Figure 6.** Coalification pattern of Pennsylvanian coal basins of the eastern United States. Data on rank refer to the Herrin (No. 6) Coal Member of the Illinois Basin and its equivalents in the other coal basins, in particular the Middle Kittanning seam of the Appalachian region. Specifications of rank follow ASTM Standard Designation D. 388-66.

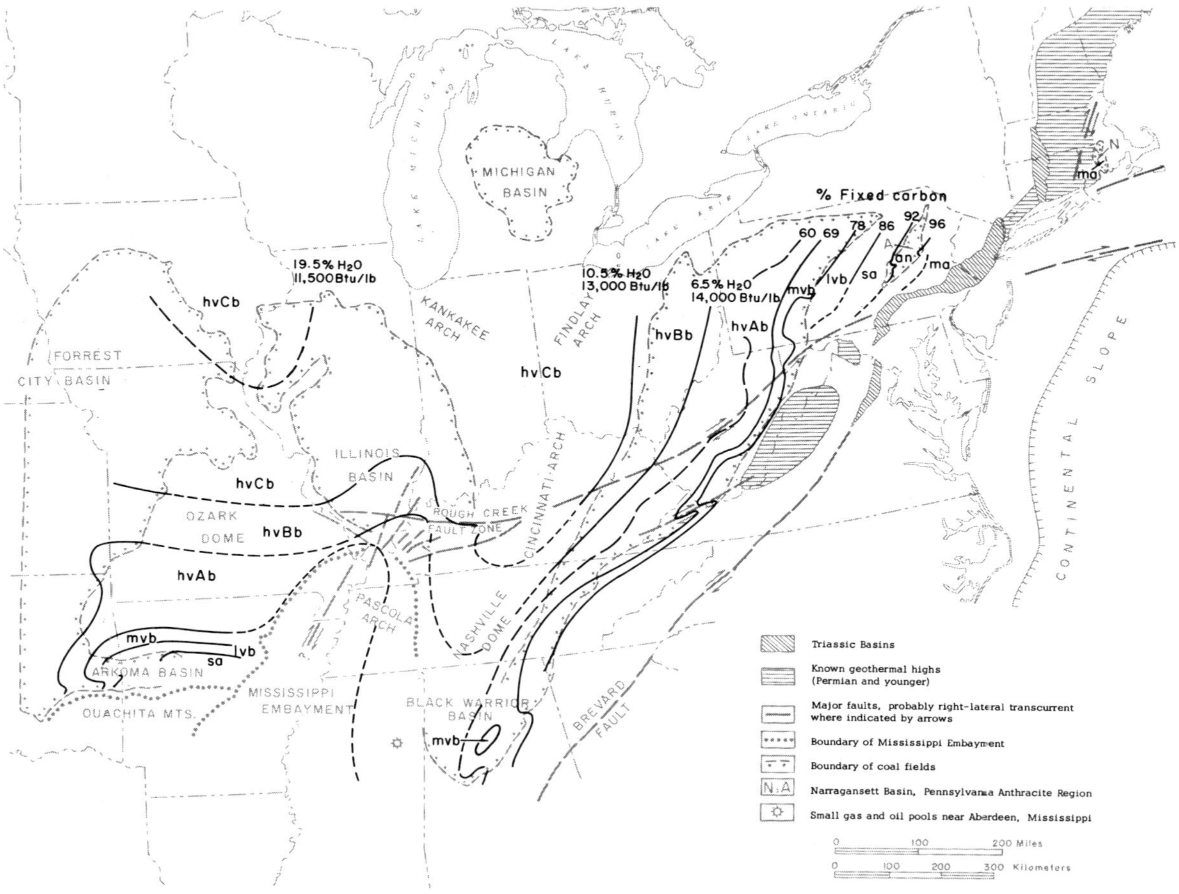

Triassic Basins

Known geothermal highs (Permian and younger)

Major faults, probably right-lateral transcurrent where indicated by arrows

Boundary of Mississippi Embayment

Boundary of coal fields

N, A Narragansett Basin, Pennsylvania Anthracite Region

Small gas and oil pools near Aberdeen, Mississippi

0 100 200 Miles

0 100 200 300 Kilometers

**Figure 7.** Coalification pattern of Pennsylvanian coal basins of the eastern United States. Data on rank are plotted regardless of stratigraphic age of coals that occur near the surface. Based on White, 1915; Fuller, 1919, 1920; Semmes, 1920, 1929; Eby, 1923; Moulton, 1925; Croneis, 1927; Campbell, 1930; Miser, 1934; Hendricks, 1935; Trumbull, 1960.

most pronounced perpendicular to the structural trend and are only minor parallel to it, as a reflection of the normal variations in sediment thickness. If this rank increase were related to northwardly increasing depth of burial, one would expect to see a corresponding thickening of Pennsylvanian stratigraphic units in the area, though the thickness patterns in younger sediments (which determine predeformational burial rank and which are, of course, not preserved) may have differed somewhat from the ones that are still preserved. Neither Branson's (1962) isopachous maps of the Pottsville, Allegheny, Conemaugh, Monongahela, and Dunkard Groups nor Wanless's (1947) map of the Lampasas Group show any such change in the area under question. In most of the maps, the isopachs follow the structural trend. In some maps, northwardly decreasing thicknesses can be observed, particularly in the lower stratigraphic units. Such lack of northward thickening of Pennsylvanian stratigraphic units suggests that the described, rather rapid, rank increase parallel to the structural trend has some cause other than increasing depth of burial to the north (see below: Isoranks as Expression of Large-Scale Geothermal Heating).

Less conspicuous, and anomalous only in comparison to other coal basins and to the situation in central and southern West Virginia, is the well-exhibited eastward rank increase of the reference seam in the bituminous coal area of Pennsylvania (Stadnichenko, 1934), from the gently deformed Plateau area into the more tightly folded Appalachian Mountains (Fig. 6).

As I pointed out earlier, in the Ruhr district of Germany, rank in the Sonnenschein coal seam decreases as the more tightly folded and intensely thrusted areas in the southeast are approached (Fig. 1). A similar pattern seems to exist in central and southern West Virginia, although, because the reference coal would be well above the present erosion level, it is not expressed in the coalification map of the Herrin-Middle Kittanning coal seams (Fig. 6). Spot checks of lower coals (in particular the detailed study of Heck, 1943) reveal, however, that the fixed carbon content of the No. 3 Pocahontas coal, for instance, decreases in a southeastward direction in McDowell County, southern West Virginia.

Fixed carbon values within any reference seam tend to reach a high in McDowell, Wyoming, Raleigh, Fayette, and Nicholas Counties in southern West Virginia and then tend to either decrease or stay at the same level farther eastward. Certainly, no rank increase within a reference seam similar to that in Pennsylvania can be detected in central and southern West Virginia and farther south. I believe that this is the normal pattern, similar to that of the Ruhr district, and that former depth of burial of any Pennsylvanian reference seam would probably decrease quite rapidly eastward into the Appalachians, where folding and subsequent uplift and erosion probably took place earlier than farther west. Information on such eastward changes in rank within reference strata in central and southern West Virgina, the most promising area for such studies, is difficult to obtain because of the complex geologic structure and the lack of coals in the area. However, since the development of methods of studying the rank of finely disseminated coaly particles in clastic sediments ("phytoclasts"; see Bostick, this volume) by measuring their reflectance under the microscope, it should now be possible to fill in such gaps in the coalification pattern of an area. At this time, however, we can only speculate.

**Isoranks as Expression of Large-Scale Geothermal Heating**

Both field and laboratory evidence and physicochemical considerations have led to the conclusion that elevated temperatures and length of exposure to elevated temperatures are the controlling factors in bringing about the chemical changes that characterize the coalification process: increased fixed carbon and carbon contents, increased heating value and reflectance, and decreased volatile matter, oxygen, and hydrogen contents. The influence of confining and directional pressure, especially on such physical properties as porosity and porosity-related parameters and on the direction and speed of chemical processes, also is important, but the geothermal history of an area determines to a high degree the maturity of coaly material in the area.

The great sensitivity of coal to even minor temperature differences persisting over a long period was first described from the Saar Basin (Damberger, 1966a, 1966b, 1968; Teichmüller and Teichmüller, 1966b, 1968b). In boreholes, change in rank with depth and horizontal change within an anticline were found to be less than normal within thick sandstone sequences, because temperature changes are less than normal in such sequences, which have greater than normal heat conductivity (Fig. 8).

The sensitivity of coaly material to prolonged heating is particularly well exhibited by the aureoles of high-rank coal around deep-seated plutonic intrusions. A few examples of such telemagmatic coalification are cited in the introduction to this paper (Figs. 4 and 5); many more have been described in the literature. As almost every coal basin has examples of telemagmatic coal metamorphism, it seems only logical to look for coalification patterns in our maps (Figs. 6 and 7) that would conform to the usual criteria of large-scale telemagmatic coalification: (1) isoranks dipping independently of beds and cutting across the regional structural fabric; (2) concentric pattern of isoranks, with anthracites and meta-anthracites often in the center; common association of rank anomalies with magnetic and (or) gravimetric anomalies; and (3) indications of hydrothermal activity (such as galena, sphalerite, fluorite, or siderite mineralizations).

Several areas in the coalification maps seem to exhibit one or more of these conditions and qualify more or less readily as telemagmatic coalification anomalies.

**Rhode Island Meta-Anthracite Region.** Zartman and others (1970) used a large number of K-Ar age determinations to delineate a Permian geothermal disturbance that extends from Long Island Sound northward into southwestern Maine (see cross-hatched area in Figs. 6 and 7). The Narragansett Basin, with its Pennsylvanian anthracites and meta-anthracites ("N" in Figs. 6 and 7), lies in the eastern portion of the geothermal anomaly. Zartman and others discussed in great detail the possible causes for the development of this late Paleozoic geothermal disturbance. The geologic history of the area is complex, and a number of causes, such as plutonic activity, major faulting, regional metamorphism, and late Paleozoic deep burial followed by uplift and erosion, could singly or in combination account for the development of the geothermal anomaly. They also speculated that the large-scale continental displacements and the associated crustal unrest that took place during the early stages of the opening of the North Atlantic at this time may provide the clue to the origin of this geothermally disturbed area. The fault zone in the Long Island area, approximately along the fortieth parallel (King, 1970, p. 96-98), that separates the Northern from the Southern Appalachians probably extended far into Africa in pre-continental drift times (Fig. 9): the Tizi n'Test Fault of Morocco is a late Paleozoic, right-lateral strike-slip fault with a possible displacement of over 200 km, characteristics that might also apply to the Long Island fault zone.

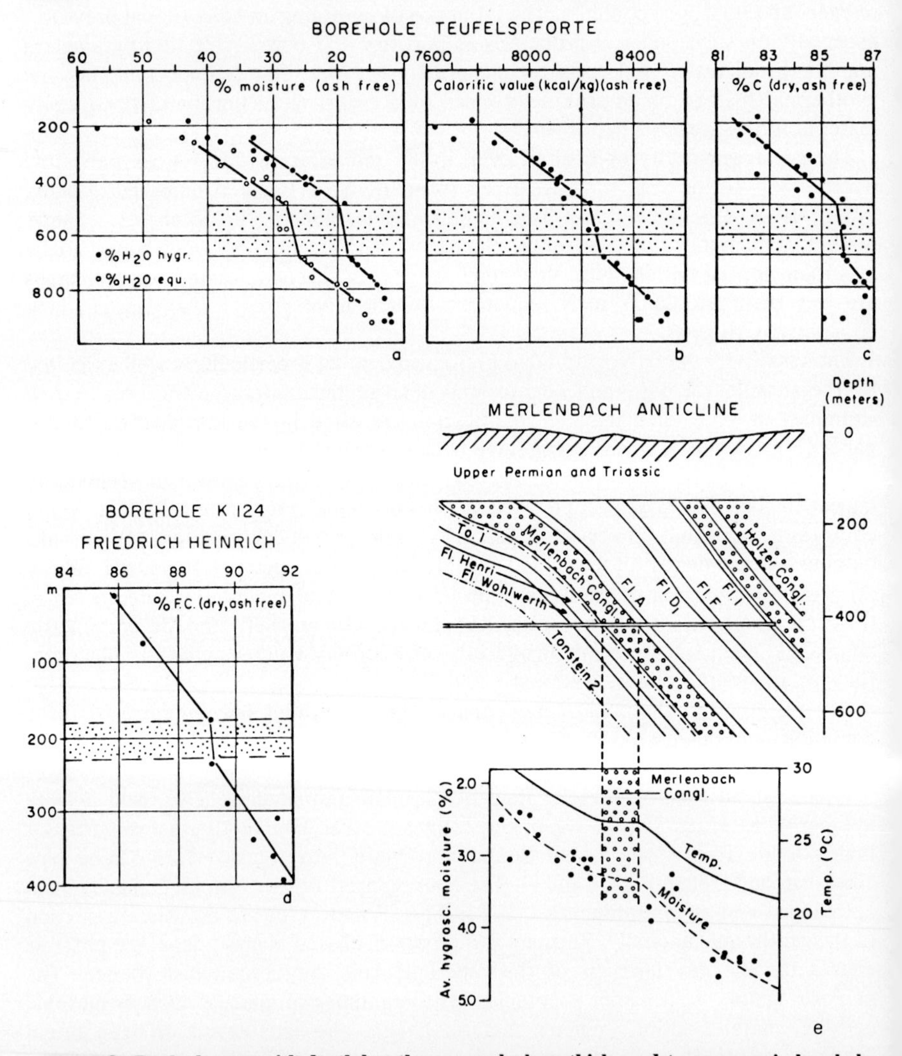


**Figure 8. Rank changes with depth less than normal where thick sandstones occur in boreholes or cross sections, as first measured in borehole Teufelspforte, Saar coal district (a through c; from Damberger, 1966b). The pattern was observed thereafter in many other cases, for example, borehole K 124 Friedrich Heinrich, Niederrhein district (d, from Teichmüller and Teichmüller, 1966b) and Merlenbach Anticline (e, from Damberger, 1968), where both present temperature and rank change only very little along the main gallery of the Warndt mine as the Merlenbach Conglomerate is crossed.**

I propose here that the anthracites and meta-anthracites of the Narragansett Basin owe their high rank to the same late Paleozoic regional heating of the area that reset the K-Ar atomic clocks when crustal unrest was prevalent during the initial stages of the opening of the North Atlantic. It is well known that rifting is associated with both large-scale strike-slip faulting (Illies, 1969) and greatly increased heat flow in the affected areas (Baker and others, 1972; Teichmüller, 1970).

**Anthracite and Low- to Medium-Volatile Bituminous Coal Fields of Pennsylvania.** K-Ar data indicate that the late Paleozoic geothermal anomaly referred to above not only extends northward from Long Island Sound, but also westward, at least as far as the northern end of the Triassic Basin of eastern Pennsylvania (horizontal hachures in Figs. 6 and 7, after Clark and Kulp, 1968, and Zartman and others, 1970). When the previously discussed unusual coalification patterns in the Anthracite Region and in the eastern portion of the bituminous coal field of Pennsylvania are considered, it appears that the thermal anomaly extends much farther west than is indicated by the K-Ar data. Coal is much more sensitive than minerals to long-range heating, and it seems quite reasonable that the areal extent of anomalously high-rank coal would be much greater than the area of anomalously low K-Ar ages. The inferred large geothermal anomaly coincides quite well with the very distinct boundary between the Northern and Southern Appalachians. It has the same trend as the east-west portion of the Triassic graben system in eastern Pennsylvania and continues the continental-scale strike-slip fault system that follows the Long Island Sound area. The fault system probably extended far into Africa before the continents became separated (Fig. 9) and also westward into the North American block, as indicated in Figures 6 and 7 (Drake and Woodward, 1963; King, 1970; Woolard and Joesting, 1965). Hydrothermal mineralization (Pb-Zn-Cu) in Pennsylvania is found mainly along a structurally disturbed zone that trends northwest from the area where the Triassic Basins cross the boundary between Pennsylvania and Maryland (Smith and others, 1971). This and a few other areas of hydrothermal mineralization would fall well within the limits of the thermal anomaly outlined by the area of anthracitic, low- and medium-volatile bituminous

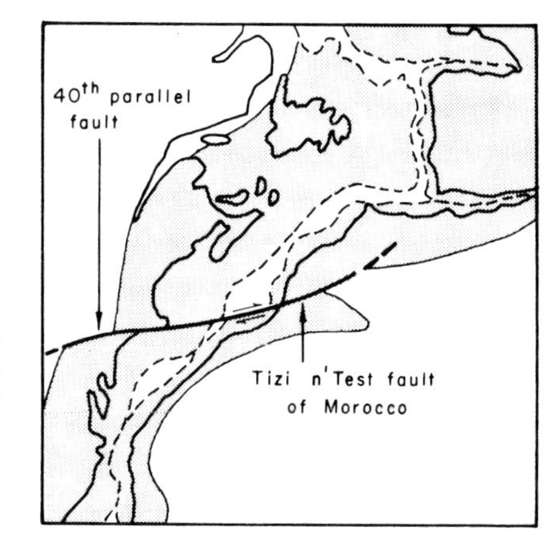


**Figure 9. Mattauer and others (1972) suggested extension of the Tizi n'Test strike-slip fault of Morocco into the North American continent in pre-continental drift times. Stippled area indicates late Paleozoic fold belt.**

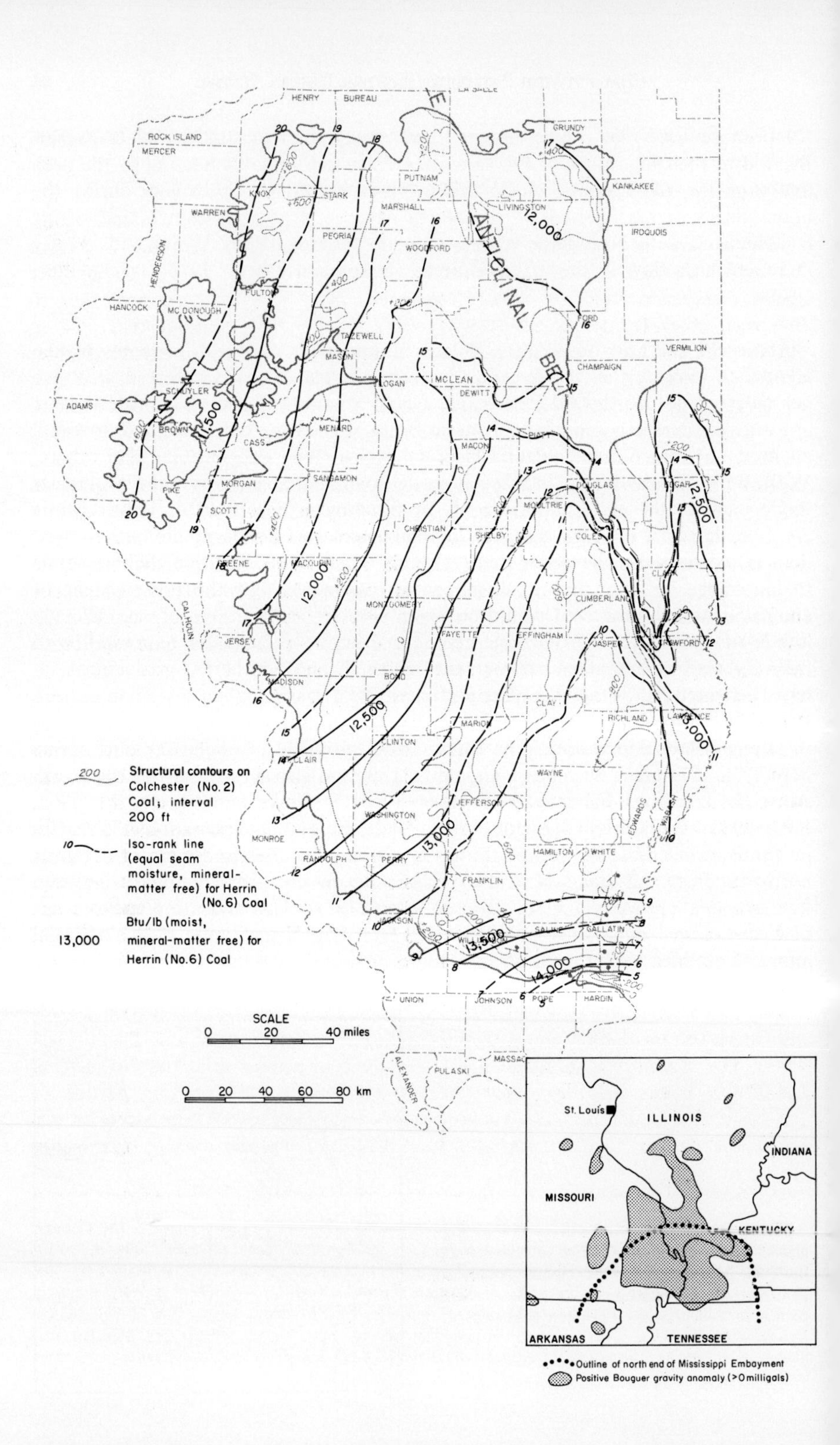

Structural contours on Colchester (No. 2) Coal; interval 200 ft
Iso-rank line (equal seam moisture, mineral-matter free) for Herrin (No. 6) Coal
13,000 Btu/lb (moist, mineral-matter free) for Herrin (No. 6) Coal
SCALE
0 20 40 miles
0 20 40 60 80 km
ANTICLINAL BELT
St. Louis
ILLINOIS
INDIANA
MISSOURI
KENTUCKY
ARKANSAS
TENNESSEE
Outline of north end of Mississippi Embayment
Positive Bouguer gravity anomaly (>0 milligals)

coals in Figure 6. Both high coal rank and mineralization might well be related to deep-seated plutonism in late Paleozoic–early Mesozoic times along the transcurrent fault system indicated in the two coalification maps (Figs. 6 and 7).

**Area of Low- and Medium-Volatile Coals in Southern West Virginia.** In Figure 7, in which the rank of coals at or near the surface regardless of their stratigraphic age is mapped, a bulge of higher rank in southern West Virginia protrudes to the west. Heck (1943), in his detailed study of coal rank of this area, pointed out that the eastward rank increase can be attributed entirely to greater former depths of burial of these successively older coal seams. He demonstrated that within reference seams, rank decreases eastward toward the more intensely folded region, probably as an indication of eastwardly decreasing former depths of burial of these reference seams. Actually, as in the southern Ruhr district (Fig. 1), the eastward decrease in rank within reference seams may formerly have been even more pronounced and may have been modified by a thermal anomaly in western Virginia and eastern West Virginia. The anomaly extends along the thirty-eighth parallel (Dennison and Johnson, 1971) from the Triassic Basin to within 20 to 30 mi of the bulge on the coalification map (horizontal hachures in Figs. 6 and 7). The igneous activity in that area is young, and the most recent activity (Eocene) seems to have been farthest west. A relation to major transcurrent basement faults with right-lateral displacements of many miles (Woodward, 1968) may well exist, as is indicated on the two coalification maps (Figs. 6 and 7).

**Arkoma Basin.** In the Arkoma Basin, rank increases to semianthracitic rather rapidly to the east from the Oklahoma-Arkansas boundary (Fig. 7). The coals are of about the same age throughout the area, and no significant change of thickness, which might help explain the rank changes, has been observed in west-east direction in the Pennsylvanian Period. Rapid change in rank parallel to the structural grain suggests the development of a geothermal anomaly in this area some time after the late Pennsylvanian Period. Igneous intrusions of Mesozoic to Cenozoic age and still-active hot springs in the eastern Ouachita Mountains and the adjacent areas of the Mississippi Embayment can be cited in support of this interpretation.

**Southern Part of the Illinois Basin.** The coalification pattern of southern Illinois also suggests anomalous conditions (Damberger, 1971; Bostick and Damberger, 1971). The isoranks in a single coal follow the structural contours in most of the Illinois Basin, but they depart from this pattern in the southern portion of the basin (Fig. 10). From the vicinity of St. Louis southward, the isoranks run independently of the trend of the structural contours, in places even at right angles

**Figure 10. Coalification pattern of Herrin (No. 6) Coal Member and its relation to the tectonic structure of the Illinois Basin (from Damberger, 1971). Only lines of equal inherent seam moisture are shown; lines of equal calorific value run parallel in positions indicated by Btu per lb values on map (see Fig. 13). Inset map shows outlines of a positive gravity anomaly in the northern Mississippi Embayment that extends into the southern portion of the Illinois Basin (after Wollard and Joesting, 1965). The gravity high might outline the position and shape of a deep-seated igneous intrusion that might have supplied heat to increase coal rank in southern Illinois.**

to them. Another anomalous situation exists along the southern margin of the Illinois Basin, where the isoranks dip more steeply toward the north than do the coal seams (Fig. 11). This would suggest that the greatest deposition in the Illinois Basin was toward the south late in Pennsylvanian and Permian times, when the main coalification took place, and that during that period both seams and isoranks dipped southward, the isoranks somewhat less than the seams, as is normal. A subsequent upwarping of the strata would explain the relation between isoranks and seams in that part of the Illinois Basin.

The trend of the isoranks more or less perpendicular to the structural contours as far north as St. Louis, however, would still be left unexplained by this sequence of events. The Bouguer gravity map of Woolard and Joesting (1965) shows a gravity high of about 100 by 70 mi, with a northward extension toward St. Louis of 20 by 60 mi (inset map in Fig. 10). This gravity anomaly could well represent the position and shape of a plutonic intrusion, which might have been a source of heat for the coals of anomalously high rank in southern Illinois. Such a possibility was mentioned by Dr. Marlies Teichmüller during the 1964 American Conference on Coal Science, after J. A. Simon had doubted that differences in former depth of burial alone could account for the observed southward increase in rank in the Illinois Basin (Schapiro and Gray, 1966, p. 216). Possibly, deeper burial *and* increased heat flow were instrumental in the development of the unusual coalification pattern in southern Illinois.

**Mississippi Embayment.** The igneous activity in both Arkansas and southern Illinois is probably closely related to the formation of the Mississippi Embayment depression. A step further would connect the two anomalously high-rank areas of Arkansas

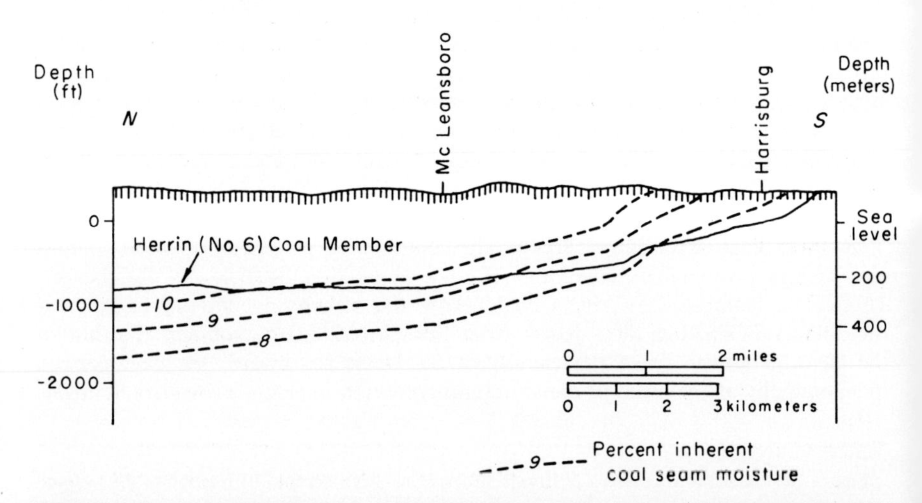


**Figure 11. In a north-south cross section through the southern portion of the Illinois Basin, the isoranks have a somewhat greater dip than the strata. This unusual coalification pattern indicates either southwardly increasing burial depth during coalification prior to the tectonic closing of the Illinois Basin in its present form or southwardly increasing telemagmatic heating (see insert map in Fig. 8) and thus change of a previously existing coalification pattern (from Damberger, 1971).**

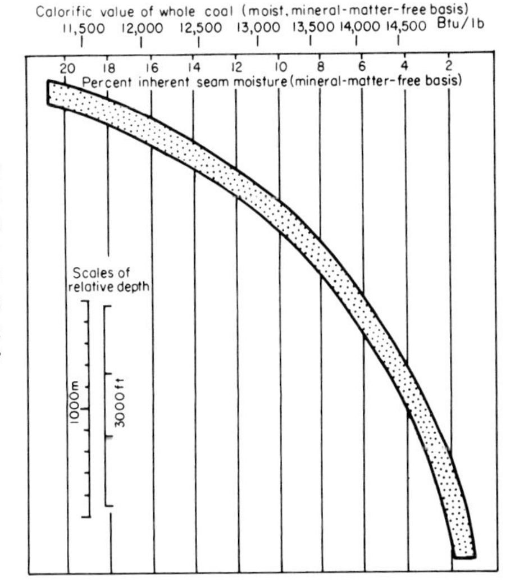


**Figure 12. Change of rank with relative depth for the Illinois Basin. Rank is expressed either by inherent seam moisture or calorific value of moist, mineral-matter-free coal (after Damberger, 1971). Rank generally changes along the trend line, with some variability as indicated by its width. The rank change shown would be observed in shafts, boreholes, or within a mine at different depths.**

and Illinois, as was tentatively done in Figure 7. Clearly, this is highly speculative. The basic assumption for the proposed configuration of high-rank areas within and around the Mississippi Embayment is the development of a geothermal anomaly in the region. The suspected geothermal anomaly in the western half of the Mississippi Embayment becomes prominent through the connection of the high-rank coal areas of the Arkoma, Illinois, and Black Warrior Basins (Fig. 7). The 69 percent fixed-carbon surface dips under the high-volatile bituminous coal areas of western Kentucky and of the Black Warrior Basin in Alabama. Its intersection with the surface of the Paleozoic strata within the Mississippi Embayment falls somewhere in the position indicated in Figure 7. The only hint for the possible position of this boundary comes from the small gas and oil pools in Paleozoic strata around Aberdeen, Mississippi (Fig. 7). Gas and condensates predominate in these pools, and oil was produced from only one small field for a short period. According to the Carbon Ratio Rule, this would indicate proximity to the 69 percent fixed-carbon line (Bostick and Damberger, 1971).

## Coalification Gradients

No special effort was made in this study to establish rank changes by depth (Hilt's rule, 1873). For the Illinois Basin, the change of rank as measured by inherent moisture (mineral-matter-free or ash-free basis) or calorific value (moist, mineral-matter-free basis) was established in a previous study (Damberger, 1971; Fig. 12). Moisture and calorific value (moist basis) are linearly related for most of the low-rank bituminous coals (Fig. 13).

For medium- and low-volatile bituminous and anthracitic coals, the fixed carbon (dry, ash-free basis) is used as a parameter of rank. Rank changes with depth in the Appalachian Region were checked for a few locations during the present study; no significant differences from the changes established in European coal

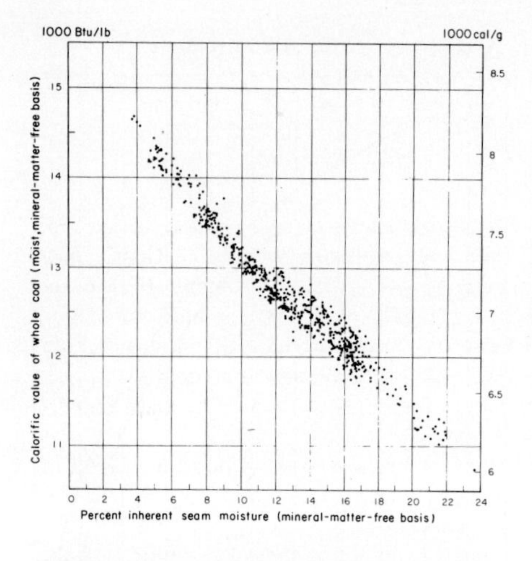


**Figure 13. Inherent seam moisture and calorific value on the moist, mineral-matter-free basis correlate well and can be used as equally suitable rank parameters for Illinois coals and other bituminous coals of the same rank. Data are from face channel samples of Illinois coals (after Damberger, 1971).**

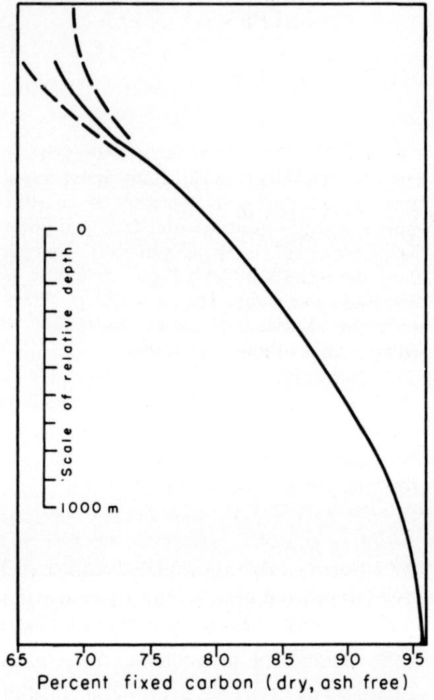


**Figure 14. Change of fixed carbon contents of dry, ash-free coal with relative depth in German coal fields (from Patteisky and others, 1962). Changes of fixed carbon with depth in Pennsylvanian coal fields of the United States seem to be of the same order of magnitude.**

fields by Patteisky and Teichmüller (1960) were noticed (Fig. 14). More detailed studies, especially of boreholes and mines with better control of depth, location, and distance between coal seams, are needed. Only if gradients are well known will it be possible to study the intricate relations between folding, faulting, and coalification in the Appalachian Region and other areas with relatively intense folding and faulting.

## ACKNOWLEDGMENTS

William A. Olsson did much of the collecting, computing to a standard basis, averaging, and plotting of the data that were used for the coalification maps (Figs. 6 and 7).

James M. Schopf, U.S. Geological Survey; Peter A. Hacquebard, Geological Survey of Canada; and members of the Coal Section of the Illinois Geological Survey read the manuscript; their critical remarks guided me in some important changes on the original manuscript.

I am also grateful to many others with whom I discussed the issues of this paper.

## APPENDIX 1. COLLECTION AND MANIPULATION OF DATA FOR COALIFICATION MAPS

The U.S. Bureau of Mines (USBM) has published thousands of coal analyses since 1953 on a yearly basis in its series, Analyses of Tipple and Delivered Samples of Coal. These samples represent the main body of data on which the two coalification maps (Figs. 6 and 7) are based. In Illinois, the extensive collection of analyses of face-channel samples at the Illinois State Geological Survey served as basis for the construction of the coalification maps (Damberger, 1971). A comparison of the rank data of the Survey face-channel samples and the USBM tipple and delivered samples that were used in constructing the maps showed good agreement.

Moisture rather than calorific value to a moist basis was used as the parameter to indicate rank in high-volatile bituminous coals (Fig. 13). Moisture was computed to the ash-free basis. Samples with excessive ash content (more than about 15 percent) were not normally used; sulfur was included in ash if it exceeded about 3 percent, but most high-sulfur samples had to be excluded because their ash contents were too high.

Moisture contents determined for fine coal tend to be higher than normal; sizes such as 0 by 1/4, 0 by 1-1/4, or 0 by 5/8 inches (USBM publications) all tend to be high in moisture (probably because of the relatively high surface area of fine material) and were often excluded from the average of any given location. Locations of the mines as reported could frequently not be determined, and data were plotted at the towns listed in the USBM publications.

Several steps of averaging the moisture values are involved in the development of the coalification maps. Annual averages were derived from mines sampled several times a year, mine averages were computed from several annual averages, and town averages were calculated from data from several mines located in an area. When the lines of approximately equal moisture contents are drawn, another step of averaging is built in: lines are generally drawn smoothly to show only the regional trend. Significant local anomalies may thus have been omitted. It is my opinion that only detailed coalification studies on the basis of samples that are taken specifically for that purpose would justify showing local, relatively small-sized anomalies in the coalification pattern.

For the construction of the single seam coalification map (Fig. 6), data from seams as far as 1,000 to 1,500 ft above and below the reference seam were used to obtain a sufficient density of rank data. But most of the data are from the reference seam itself, or from seams only a few tens of feet away from it. For high-volatile bituminous coals, moisture contents were converted along the curve of Figure 12. For higher rank coals in which fixed carbon (dry, ash free) is used as rank parameter, the curve in Figure 14 was used for the conversion. The accuracy of such conversions decreases with the distance of the data coal seam from the reference seam because the gradients change regionally for various reasons, and the distance between the seams may not be accurately known. Also, the general applicability of the two charts has not yet been established by sufficient checking. However, I think that the coalification pattern of Figure 6 gives a fair representation of the regional changes in rank of the Herrin–Middle Kittanning seams. Many areas had to be left blank; other seams might be used as reference seams more advantageously in those regions.

In the high-rank bituminous coal fields (medium-volatile bituminous to meta-anthracite), I relied mainly on the many previous coalification maps published by several individuals (White, 1915; Fuller, 1919, 1920; Semmes, 1920, 1929; Eby, 1923; Moulton, 1925; Croneis, 1927; Campbell, 1930; Miser, 1934; Hendricks, 1935; Trumbull, 1960), on those by the Geological Surveys of the states that are covered in my maps, and especially on the map by Stadnichenko (1934) for Figure 6.

## REFERENCES CITED


Baker, B. H., Mohr, P. A., and Williams, L. A. J., 1972, Geology of the Eastern Rift System of Africa: Geol. Soc. America Spec. Paper 136, 67 p.

Bostick, N. H., and Damberger, H. H., 1971, The carbon ratio rule and petroleum potential in NPC region 9: Illinois Geol. Survey Illinois Petroleum 95, p. 142-151.

Branson, C. C., 1962, Pennsylvanian System of central Appalachians, *in* Branson, C. C., ed., Pennsylvanian System in the United States: Am. Assoc. Petroleum Geologists, p. 97-116.

Campbell, M. R., 1930, Coal as a recorder of incipient rock metamorphism: Econ. Geology, v. 25, no. 7, p. 675-696.

Clark, G. S., and Kulp, J. L., 1968, Isotopic age study of metamorphism and intrusion in western Connecticut and southeastern New York: Am. Jour. Sci., v. 266, no. 10, p. 865-894.

Croneis, C., 1927, Oil and gas possibilities in the Arkansas Ozarks: Am. Assoc. Petroleum Geologists Bull., v. 11, no. 3, p. 279-297.

Damberger, H. H., 1966a, Die Abhängigkeit des Inkohlungsgradienten vom Gesteinsaufbau: Zeitschr. deutsch. geol. Gesell. Jahrg. 1965, v. 117, no. 1, p. 8.

——1966b, Inkohlungsmerkmale, ihre statistische Bewertung und ihre Anwendbarkeit bei der tektonischen Analyse im saarländischen Steinkohlengebirge: Dissertation Bergakademie Clausthal, 99 p.

——1968, Ein Nachweis der Abhängigkeit der Inkohlung von der Temperatur: Brennstoff-Chemie, v. 49, no. 3, p. 73-77.

——1971, Coalification pattern of the Illinois Basin: Econ. Geology, v. 66, no. 3, p. 488-494.

Damberger, H. H., Kneuper, G., Teichmüller, M., and Teichmüller, R., 1964, Das Inkohlungsbild des Saarkarbons: Glückauf, v. 100, no. 4, p. 209-217.

Dennison, J. M., and Johnson, R. W., 1971, Tertiary intrusions and associated phenomena near the thirty-eighth parallel fracture zone in Virginia and West Virginia: Geol. Soc. America Bull., v. 82, no. 2, p. 501-508.

Drake, C. L., and Woodward, H. P., 1963, Appalachian curvature, wrench faulting, and offshore structures: New York Acad. Sci. Trans., Ser. 2, v. 26, p. 48-63.

Eby, J. B., 1923, The possibilities of oil and gas in southwest Virginia as inferred from isocarbs: Am. Assoc. Petroleum Geologists Bull., v. 7, no. 4, p. 421-426.

Edmunds, W. E., 1971, Pennsylvania. Bituminous and anthracite coal fields, *in* 1971 Keystone coal industry manual: New York, McGraw-Hill Book Co., p. 443-454.

Fuller, M. L., 1919, Relation of oil to carbon ratios of Pennsylvanian coals in north Texas: Econ. Geology, v. 14, no. 7, p. 536-542.

——1920, Carbon ratios of Carboniferous coals of Oklahoma and their relation to petroleum: Econ. Geology, v. 15, no. 3, p. 225-235.

Heck, E. T., 1943, Regional metamorphism of coal in southeastern West Virginia: Am. Assoc. Petroleum Geologists Bull., v. 27, no. 9, p. 1194-1227.

Hendricks, T. A., 1935, Carbon ratios in part of Arkansas-Oklahoma coal field: Am. Assoc. Petroleum Geologists Bull., v. 19, no. 7, p. 937-947.

Hilt, Carl, 1873, Die Beziehungen zwischen der Zusammensetzung und den technischen Eigenschaften der Kohle: Sitzungs-Ber. Aachener Ver. VDI, Heft 4, p. 194-202.

Illies, J. H., 1969, Intercontinental belt of the world rift system: Tectonophysics, v. 8, no. 1, p. 5-29.

King, P. P., 1970, Tectonics and geophysics of eastern North America, *in* Johnson, H., and Smith, B. L., eds., The megatectonics of continents and oceans: New Brunswick, Rutgers Univ. Press, p. 74-112.

Mattauer, M., Proust, F., and Tapponier, P., 1972, Major strike-slip fault of late Hercynian age in Morocco: Nature, v. 237, no. 5360, p. 160-162.
Miser, H. D., 1934, Relation of Ouachita Belt of Paleozoic rocks to oil and gas fields of mid-continent region: Am. Assoc. Petroleum Geologists Bull., v. 18, no. 8, p. 1059-1077.
Moore, R. C., chm., 1944, Correlation of Pennsylvanian formations of North America: Geol. Soc. America Bull., v. 55, no. 6, p. 657-706.
Moulton, G. F., 1925, Carbon ratios and petroleum in Illinois: Illinois Geol. Survey Rept. Inv. no. 4, 18 p.
Patteisky, K., and Teichmüller, M., 1960, Inkohlungs-Verlauf, Inkohlungs-Masstäbe und Klassifikation der Kohlen auf Grund von Vitritanalysen: Brennstoff-Chemie, v. 41, no. 3, p. 79-84; no. 4, p. 97-104; no. 5, p. 133-137.
Patteisky, K., Teichmüller, M., and Teichmüller, R., 1962, Das Inkohlungsbild des Steinkohlengebirges an Rhein und Ruhr, dargestellt im Niveau von Flöz Sonnenschein: Fortschr. Geologie Rheinland u. Westfalen, v. 3, no. 2, p. 687-700.
Schapiro, N., and Gray, R. J., 1966, Physical variations in highly metamorphosed Antarctic coals, *in* Gould, R. F., ed., Coal Science: Washington, D.C., Advances in Chemistry Ser., Am. Chem. Soc., p. 196-217.
Semmes, D. R., 1920, Petroleum possibilities of Alabama: Pt. I, northern Alabama: Alabama Geol. Survey Bull. 22, p. 54-57.
——1929, Oil and gas in Alabama: Alabama Geol. Survey Spec. Rept. 15, p. 24-26, 61-64.
Skipsey, E., 1959, Rank variations in the coal seams of north-east Stirlingshire: Inst. Mining Engineers Trans., v. 119, pt. 1, p. 23-36.
Smith II, R. C., Herrick, D. C., Rose, A. R., and McNeal, J. M., 1971, Zinc-lead occurrences near Mapleton, Huntington Co., Pennsylvania: Pennsylvania State Univ. Mineral Conserv. Ser., Circ. 83, 37 p.
Stadnichenko, Taisia, 1934, Progressive regional metamorphism of the lower Kittanning coal bed of western Pennsylvania: Econ. Geology, v. 29, no. 6, p. 511-543.
Teichmüller, M., 1970, Bestimmung des Inkohlungsgrads von kohligen Einschlüssen in Sedimenten des Oberrheingrabens—ein Hilfsmittel zur Klärung geothermischer Fragen: Graben Problems, Internat. Upper Mantle Proj. Sci. Rept., no. 27, p. 124-142.
Teichmüller, M., and Teichmüller, R., 1966a, Die Inkohlung im Saar-Lothringischen Karbon, verglichen mit der im Ruhrkarbon: Zeitschr. deutsch. geol. Gesell, Jahrg. 1965, v. 117, no. 1, p. 243-279.
——1966b, Inkohlungsuntersuchungen im Dienst der angewandten Geologie: Freiberger Forschungshefte, C 210 Sonderveranstaltungen, p. 155-195.
——1968a, Cainozoic and Mesozoic coal deposits of Germany, *in* Murchison, D., and Westoll, T. S., eds., Coal and coal-bearing strata: Edinburgh, Oliver and Boyd, p. 347-379.
——1968b, Geological aspects of coal metamorphism, *in* Murchison D., and Westoll, T. S., eds.: Coal and coal-bearing strata, Edinburgh, Oliver and Boyd, p. 233-267.
Trumbull, James, 1960, Coal fields of the United States, Sheet 1: U.S. Geol. Survey Map, scale 1: 5,000,000.
Wanless, H. R., 1947, Regional variations in Pennsylvanian lithology: Jour. Geology, v. 55, no. 3, p. 237-253.
White, C. D., 1915, Some relations in the origin between coal and petroleum: Washington Acad. Sci. Jour., v. 5, no. 6, p. 189-212.
Wood, G. H., Trexler, J. P., and Kehn, T. M., 1969, Geology of the west-central part of the southern anthracite field and adjoining areas. Pennsylvania: U.S. Geol. Survey Prof. Paper 602, 150 p.
Woodward, H. P., 1968, Basement map: 1968 Geologic map of West Virginia: West Virginia Geol. and Econ. Survey Geol. Map.

Woolard, G. P., and Joesting, H. R., 1965, Bouguer gravity anomaly map of the United States: Arlington, Virginia, Am. Geophys. Union and U.S. Geol. Survey.
Zartman, R. E., Hurley, P. M., Krueger, H. W., and Giletti, B. J., 1970, A Permian disturbance of K-Ar radiometric ages in New England: Its occurrence and cause: Geol. Soc. America Bull., v. 81, no. 11, p. 3359-3374.

SYMPOSIUM HELD AT G.S.A. ANNUAL MEETING IN MILWAUKEE, NOVEMBER 1971
MANUSCRIPT RECEIVED BY THE SOCIETY MARCH 9, 1973

Printed in U.S.A.

Geological Society of America
Special Paper 153


# Rank Studies of Coals in the Rocky Mountains and Inner Foothills Belt Canada

P. A. Hacquebard

*Geological Survey of Canada*
*Institute of Sedimentary and Petroleum Geology*
*3303 33rd Street, N.W.*
*Calgary, Alberta, Canada*

J. R. Donaldson

*Department of Energy, Mines and Resources*
*Mines Branch*
*Fuels Research Centre*
*555 Booth Street*
*Ottawa, Ontario, Canada*

**ABSTRACT**

Discussed in this study are regional and stratigraphic variations in the rank of coals in the Rocky Mountain region, as determined from vitrinite reflectance measurements. For the regional changes, the Kootenay coals of the Crowsnest Pass area have been examined; they show a progressive westward increase in rank. The variations in rank with stratigraphic position are illustrated in ten coal-bearing sections of Jurassic-Cretaceous age, situated between the Crowsnest field in the south and the Peace River canyon in the north. Both studies indicate preorogenic coalification, because the rank increases regularly with stratigraphic position, but not with geologic age, depth of mining, or degree of tectonic disturbance.

For each of the ten curves plotted, the coalification gradient is calculated in terms of percent reflectance (Ro) change per 100 m increase in depth. By relating this gradient to that of a known curve (the Peel curve of the Netherlands) a reference for comparison is obtained, which is expressed as the Peel rank ratio. Different ratios were obtained, which probably are related to variations in the temperature gradients. The lowest ratio was found in the Peace River canyon area, and the highest occurs in the Canmore coalfield.

The coalification gradient affects the availability of coking coals of most favorable rank, that is, the medium-volatile coals. With a low gradient (and corresponding steep curve), medium-volatile coals occur over a greater stratigraphic interval, with the possibility of a larger number of seams, than with a high gradient. Within limited areas of the same coalfield, the rank as determined from vitrinite reflectance can be used for correlating coal seams, provided a high coalification gradient is present. This method has been employed successfully in the Canmore coalfield on seams that lie not less than 120 ft apart stratigraphically.

## INTRODUCTION

Bituminous coals of excellent coking quality occur in a narrow zone on the eastern side of the Rocky Mountains. This zone, which is located mainly in the Inner Foothills Belt, extends for some 600 mi from the U.S. border in the south to the Peace River canyon in the north. In the southern part, it includes the coals of the Upper Elk Valley and Fernie Basin, which actually lie within the mountain ranges. The rank of coals in this zone is predominantly medium- and low-volatile bituminous, but in certain areas extends into the high-volatile range at the one extreme and into semianthracitic range on the other. The region, therefore, provides an excellent opportunity for a comparative rank study of high-rank bituminous coals. Although the over-all rank changes are generally well known, it was not until recently that more detailed rank studies on a regional basis have been undertaken. Norris (1971) examined the variations in rank of the Kootenay coals in the southern part of the region and clearly illustrated the relation between coalification and geological conditions. Such rank studies and the one presented here may be considered highly opportune at the present time, because of the greatly expanded mining activities. They can throw more light on the actual causes of coal metamorphism, predict rank changes in newly developed areas, or assist with coal exploration. Furthermore, a detailed knowledge of the variations in rank as well as petrographic composition is essential for the evaluation of coking characteristics of the coals present in a mining operation.

## RANK PARAMETER EMPLOYED

The coals of the Rocky Mountain region belong to the bituminous and anthracitic varieties, which are classified by rank according to percentage of volatile matter and calorific value. Since these parameters are determined on the entire coal substance, a chemical rank designation normally represents an aggregate value of the degree of coalification of the various constituents that are present. Within one coal, there are substantial differences in rank between these constituents, and the aggregate value is therefore controlled by the relative proportions in which they occur. For instance, in a high-volatile coal, the vitrinite component may have

32 percent volatile matter, while fusinite has only 20 percent. These differences become less as the over-all rank of the coal increases and virtually disappear in the anthracites. Precise rank evaluations based on chemical determinations therefore require a large number of analyses, unless they are made on one coal constituent only.

The most suitable constituent is vitrinite, because it is the most common and most abundant of all coal components. However, proximate analyses of vitrinite are normally not available, and tedious hand picking is required in order to collect pure samples. A better and much simpler method is provided by using the reflectance of vitrinite as a rank parameter. This is particularly so in the bituminous and anthracitic coals, where an excellent correlation exists between the volatile matter content of vitrinite and its reflectance. This correlation has been established by Kötter (1960) on the basis of numerous analyses, and his curve has been used in this study for the conversions of reflectance measurements to volatile matter that are shown in the diagrams (see Fig. 1).

Throughout the report, the vitrinite reflectance has been employed. With this

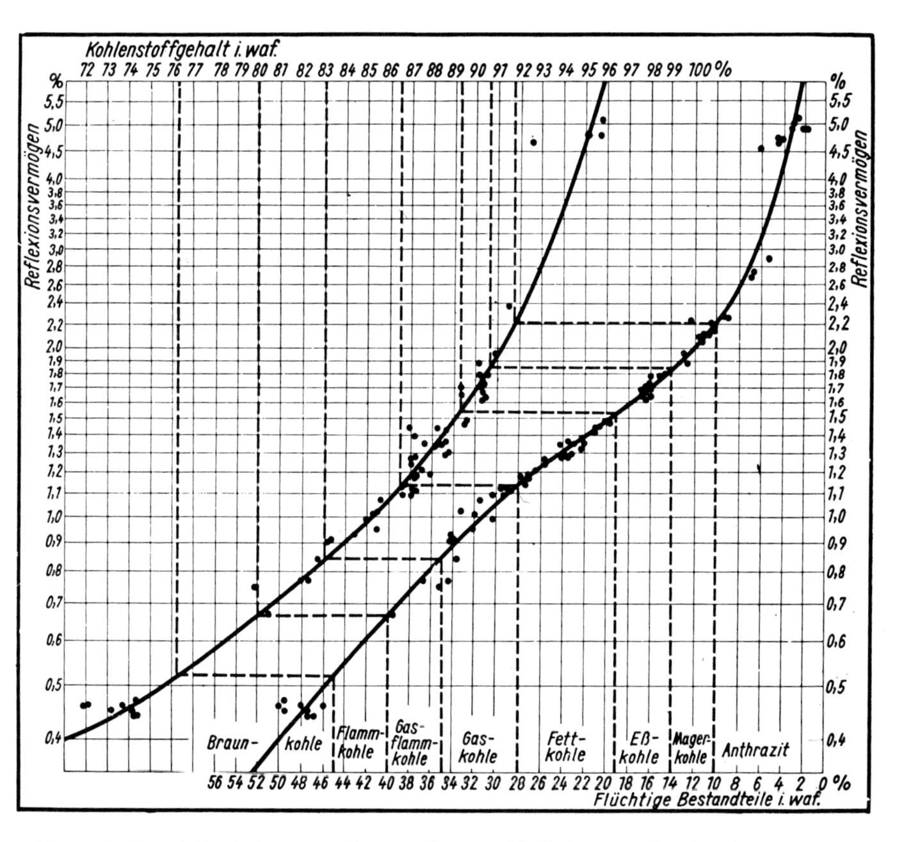


**Figure 1. Correlation between maximum reflectance (Reflexionsvermögen) and volatile matter (Flüchtige Bestandteile) and carbon content (Kohlenstoffgehalt) of vitrinite (after Kötter, 1960).**

method, reliable rank data have been obtained on representative seam samples and random collections from old mine dumps.

The technique of measuring the reflectance of polished surfaces of coal and the required equipment for this have been described by the authors in a previous publication (Hacquebard and Donaldson, 1970). Suffice it here to state that the mean of the maximum reflectance of 50 separate grains with pure vitrinite (from vitrite bands) has been used in every case to obtain the rank.

Briefly stated, the advantages of the reflectance method are as follows:

1. The rank is always determined on the same constituent, namely on vitrinite.
2. The rank can be determined on oxidized coals also, because oxidized particles can be recognized microscopically and the measurements restricted to the nonoxidized grains. This is most important when determining the rank of coal form outcrops or old mine dumps.
3. The ash content of the coal need not be considered, because no adjustments in rank to the ash-free basis are necessary.

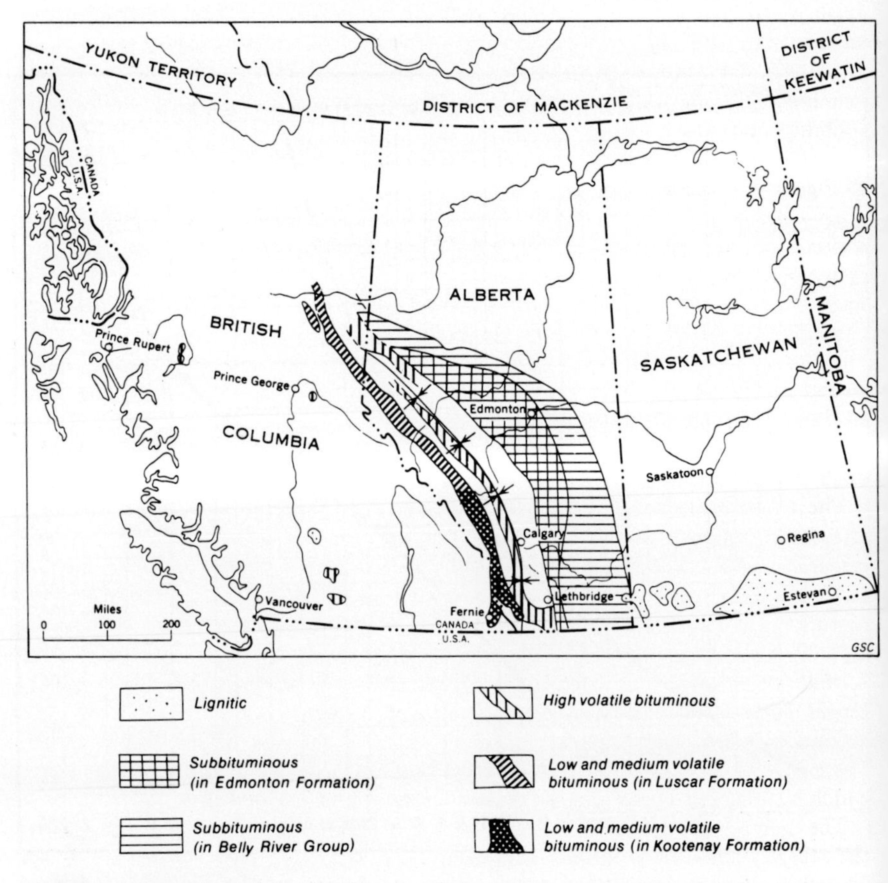


**Figure 2. General areal distribution of coals by rank in western Canada (after Latour and Chrismas 1970, modified).**

4. Even on minute particles, such as occur dispersed in clastic rocks, the degree of organic metamorphism can be measured, which is a significant factor for the occurrence of oil and gas.

## REGIONAL RANK CHANGES AND EFFECT OF TECTONISM

In western Canada, there is a progressive increase in the rank of coal from the plains region toward the Rocky Mountains (see Fig. 2). At first glance, this east-west increase seems to be related entirely to the geological age of the deposits. The lignites of Saskatchewan are Tertiary in age, the subbituminous coals of central Alberta are Upper Cretaceous, and the low- and medium-volatile bituminous coals of the Foothills and Mountain regions are Lower Cretaceous and partly Upper Jurassic in age.

However, closer inspection shows that younger coals situated in the axial region of the Alberta syncline are considerably higher in rank than the older coals that occur on the eastern flank. The high-volatile bituminous coals of the axial region are partly of Tertiary age, whereas the subbituminous coals of the Alberta Plains are Upper Cretaceous. In the latter, the older Belly River coals, moreover, are mainly subbituminous C, whereas the younger Edmonton coals are predominantly subbituminous B.

Within each formation, the rank also increases toward the west. This is illustrated in Figure 3, which shows the changes in rank in the Crowsnest Pass region of coals belonging to the Kootenay Formation and the Belly River Group. In this region, coal mining has been carried out since 1899, and as a result, there are many old adits with mine dumps from which samples can still be collected. On these samples, the rank has been determined from the vitrinite reflectance of nonoxidized particles (see Table 1). The reflectance values have been grouped together in different V-types, which are shown by the circular symbols in the legend of Figure 3. The higher the V-type, the higher is the rank of the coal. V-7 and V-9 refer to high-volatile coals with from 36 to 39 and from 31 to 34 percent volatile matter, respectively, while V-10 and V-11 are medium-volatile coals that have from 29 to 31 and from 26 to 28 percent volatile matter.

The Crowsnest coal area of Alberta lies in the Foothills Belt, which has been affected by intense folding and thrust faulting. As can be seen on the map and structure section of Figure 3, the Kootenay Formation is repeated in a number of thrust segments and on the flanks of synclines and anticlines. A major tectonic feature is the Lewis thrust fault, which borders the Rocky Mountains on the east side. This thrust has placed Paleozoic sediments on top of Mesozoic coal-bearing formations which belong to the Upper Cretaceous Belly River Group and the (partly) Upper Jurassic Kootenay Formation. This is strikingly revealed by Crowsnest Mountain, which stands as an erosional klippe 2 mi east of the Lewis thrust. The eastward movement along this thrust is of the order of 25 mi (MacKay, 1933), which is indicated by the gap in the bottom section of Figure 3.

The Belly River coal south of Sentinel (17, Fig. 3) is perfectly located for examining the relation between rank and tectonism. This coal lies within half a mile of the present fault scarp of the Lewis thrust, but formerly occurred below it as is indicated by the alignment of Crowsnest Mountain with the position of the fault farther south. The coal, therefore, must have been subjected to tremendous tectonic forces when overidden by older Paleozoic and Mesozoic rocks that were at least 2,000 ft thick (as is presently the case in Crowsnest Mountain), but more likely had obtained a total thickness of some 15,000 ft as measured from the Upper Devonian to the top of the Lower Cretaceous. However, notwithstanding this great pressure, the rank of the coal is surprisingly low, namely V-7. The nearest other Belly River coal is found at Lundbreck, some 20 mi farther east. This coal also has a rank of V-7, although here there has been but little tectonic deformation.

No tectonic effect on rank has been observed in the Kootenay coals either. At Frank, where Paleozoic sediments of the Blairmore Range are thrust over the Kootenay Formation, the rank is V-10, whereas at Grassy Mountain (5, Fig. 3) in a normal stratigraphic succession, the rank is equal to V-11. The same rank

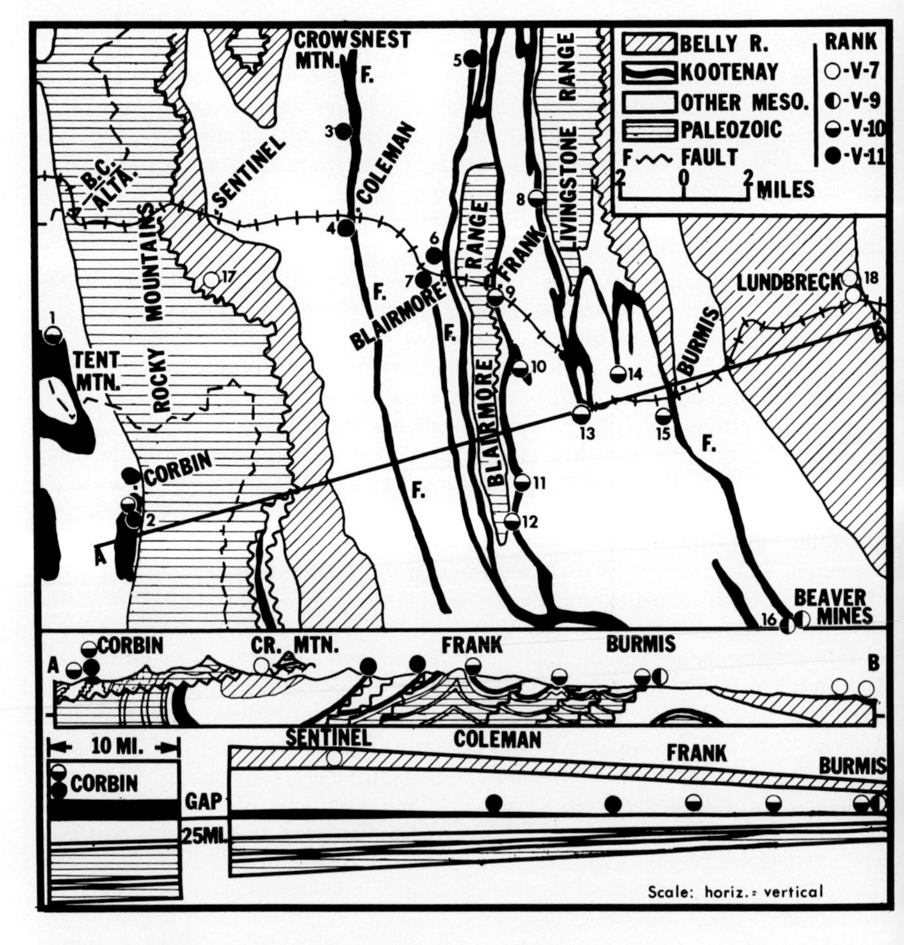


**Figure 3. Regional coal rank changes in Crowsnest Pass area shown by vitrinite reflectance types (map with pre- and post-Laramide sections after MacKay, 1933).**

Table 1. Record of Coal Samples Used in Regional Rank Study of Crowsnest Pass Area

| Index no. on Figure 3 | Sample no. | Obtained from | Location | Percent $Ro^{max}$ |
|---|---|---|---|---|
| | | Kootenay Formation | | |
| 1 | Tm-4 | Strip mine | Tent Mountain, No. 4 seam | 1.06 |
| 1 | Tm-1 | do. | Tent Mountain, No. 3 seam | 1.05 |
| 1 | Tm-2 | do. | Tent Mountain, No. 2 seam | 1.03 |
| 1 | Tm-3 | do. | Tent Mountain, No. 1 seam | 1.02 |
| 2 | 7768 | Underground mine | Corbin, Upper seam } Ro from fuel ratio | 1.06 |
| 2 | 7765 | do. | Corbin, Mammoth seam } Ro from fuel ratio | 1.14 |
| 3 | WC-II | do. | McGillivray Mine, No. 2 seam | 1.09 |
| 4 | WC-IA | do. | International mine, No. 2 seam | 1.15 |
| 5 | CQ 205 | Outcrop | Grassy Mountain 28 ft seam at base of Mutz member | 1.17 |
| 5 | 71-4 | do. | Grassy Mountain, 13 ft seam at top of Adanac member | 1.11 |
| 5 | 71-5 | do. | Grassy Mountain, 9 ft seam at base of Adanac member | 1.17 |
| 6 | WC-XVI | Underground mine | Greenhill Mine, No. 2 seam | 1.24 |
| 7 | CL-3 | Old mine dump | C & T Sunburst Mine, Blairmore | 1.12 |
| 8 | CL-12 | do. | Adit on Gold Creek, 3 mi N. of Frank | 1.09 |
| 9 | CL-4 | do. | Franco-Canadian Collieries Mine at Frank | 1.03 |
| 10 | CL-5 | do. | Hillcrest Collieries Mine at Hillcrest Mines | 1.09 |
| 11 | CL-7 | do. | Adit on Bryon Creek, 3 mi S. of CL-5 | 1.00 |
| 12 | CL-8 | Strip mine | Hastings Ridge Adanac Mine, 2.5 mi SSW of CL-7 | 1.02 |
| 13 | CL-6 | Old mine dump | Byron Creek Collieries Mine, 1.5 mi S. of Bellevue | 1.04 |
| 14 | CL-10 | do. | Leitch Collieries Mine, 1.5 mi E. of Bellevue | 1.07 |
| 15 | CL-9 | do. | Mine, 1.2 mi SSE of Burmis | 1.08 |
| 16 | BM-1 | do. | Winonis Mine at Beaver Mines | 0.94 |
| 16 | BM-2 | do. | Adit on E. side of Beaver Mines Creek | 0.91 |
| | | Belly River Formation | | |
| 17 | CL-2 | Old mine dump | Canadian-American Coal Co. Mine, 1.8 mi S. of Sentinel | 0.78 |
| 18 | Ld-1 | do. | Shaft, N. side of Crowsnest River, 0.7 mi NNW of Lundbreck | 0.76 |
| 18 | Ld-3 | do. | Shaft, S. side of Crowsnest River, 0.5 mi NNW of Lundbreck | 0.76 |

was obtained on the No. 2 seam near Coleman (4, Fig. 3), although it occurs no more than 400 ft stratigraphically from the Coleman thrust fault (Norris, 1971).

Tectonism has therefore not significantly influenced the rank of the coal; at least it has not contributed to the chemical coalification. Physical changes are undoubtedly effected, but these are mainly an increase in density and loss in moisture content, two characteristics which play an important role in the metamorphic upgrading of lignites and subbituminous coals. In bituminous coals, such changes are largely completed, and tectonic influence on these coals in terms of chemical alterations can be observed only in the immediate vicinity of thrust faults. Teichmüller and Teichmüller (1966) have shown that in the Ruhr coals this influence is restricted to a zone that is only a few hundred feet wide and is manifested by a slight change in the content of volatile matter. Tectonism, however, has a pronounced effect on the structural fabric of coal seams. Norris (1971) has shown that it causes the destruction of the primary depositional fabric by interbed slip and that this occurs anywhere in a thrust plate regardless of the proximity of faults.

Figure 3 shows that there is a progressive increase in rank of the Kootenay coals from east to west, namely, from V-9 at Beaver Mines, to V-10 between Burmis and the Blairmore Range, to V-11 west of this range and in the Coleman area. As has been pointed out, this increase toward the Rocky Mountains is unrelated to the tectonic forces that created the mountains. Not tectonism, but the initial depth of overburden that existed before the folding appears to be the controlling factor. The schematic palinspastic section at the base of Figure 3 shows that the Cretaceous formations all increase in thickness from east to west. In the Kootenay Formation, the thickness increases from about 400 ft or less in the Crowsnest Pass region to a maximum of 3,800 ft at Coal Creek, which is 14 mi west of Corbin (Norris, 1959; Jansa, 1971). The overlying Blairmore Group gains even more in thickness, namely, from about 1,500 ft on the eastern side of the study area (at Mill Creek), to 2,100 ft at Ma Butte (10 mi north of Coleman), to a maximum of 6,500 ft in the Fernie synclinorium (at Flathead River) (Norris, 1964).

Since the coals east of the Lewis thrust fault probably all lie at approximately the same stratigraphic horizon (200 to 300 ft above the base of Kootenay Formation), the observed westward increase in rank should be accompanied by a comparable increase in thickness of pre-Laramide strata above these coals. Between Beaver Mines (No. 16) and Blairmore (No. 7), the increase in overburden is represented mainly by a 500- to 600-ft thickening of strata belonging to the Lower Cretaceous Blairmore Group (Norris, 1964); the Upper Cretaceous changes but little within this distance. These 500- to 600-ft additional strata compare with a rank difference of 0.18 percent Ro that exists between samples 7 and 16, or a coalification gradient of 0.100 to 0.120 percent Ro per 100 m. These gradients compare with those in adjacent areas, which vary between 0.077 and 0.164 (sections 6 through 10, Fig. 5) and were obtained on vertical coal sequences.

From the aforementioned correspondence between vertical and regional changes in rank, a close relation with the initial, or pre-orogenic, depth of overburden is indicated. This conclusion was reached also by Norris (1971), who plotted a pre-Laramide palinspastic map in which the isopachs of the Kootenay Formation and isograds (lines of equal rank) on the basal coals run subparallel, at least in the Crowsnest coal area.

The observed westerly increase in rank also supports the view of Jansa (1971) that the coals at Grassy Mountain (No. 5) can be correlated with the basal Kootenay coals of the Fernie Basin at Michel, 16 mi west of Coleman. The former have a rank of V-11 and the latter V-14, while the upper coals of the Michel section have ranks of V-9, V-10, and V-11.

## RANK-DEPTH RELATIONS

### Effect of the Depth of Mining

In the preceding discussion, it has been shown that the rank of the coals in the Foothills Belt is of pre-orogenic origin. Little tectonic influence has been noted, and the effect of postdeformational depth of burial also is not apparent. This is clearly revealed by the curves plotted in Figure 4. They show that within individual seams the rank does not change with an increase in the depth of mining, whether the coal is low-volatile bituminous or semianthracite. This result is significant for mining, because it means that the rank of a seam as obtained at the outcrops will be the same when extracted at depth. This constancy of rank within the same seam is not always the case in coal mining; in Nova Scotia the opposite is true. Here a rank increase corresponding to a loss in volatile matter from 36 percent at the surface to 26 percent at a vertical depth of 1,240 m was observed in the Springhill No. 2 seam (Hacquebard and Donaldson, 1970).

### Coalification Curve of the Peel Region of the Netherlands

In pre-orogenic coalification, the rank changes with stratigraphic position, because the temperature at depth is related to the original thickness of overburden that existed before folding. This relation is known as Hilt's law, which was first formulated in Germany in 1873. When studying rank-depth changes in different areas, different coalification gradients are encountered which, moreover, vary with the rank increments that are involved. A different gradient occurs in the high-volatile B and C bituminous coals as compared to the medium- and low-volatile bituminous coals. In order to make comparisons, it is necessary, therefore, to have a reference curve of a stratigraphic sequence encompassing a large rank interval. Such a curve is represented in part by the coalification series encountered in wells and a deep shaft in the Peel horst of the Netherlands.

The depth-volatile relations of the (Carboniferous) coals of the Peel sequence have been published by Kuyl and Patijn (1961). They range from 33 to 8 percent in volatile matter over a depth of 1,400 m. This curve is reproduced in Figure 5, but instead of plotting the volatile matter in equal distances on the abscissa, the corresponding vitrinite reflectance values (percent Ro) have been so marked. With this procedure, a straight line relation is revealed between reflectance and stratigraphic depth in the range of high-volatile coal to the middle of semianthracite. Above and below these ranks, the curve turns rapidly and becomes asymptotic. When the relation is plotted with equal distances of volatile matter, as was done

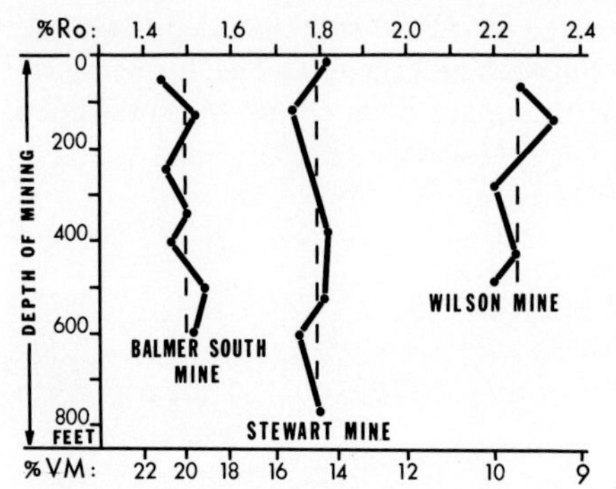


**Figure 4. Changes in rank of individual coals with depth of mining in three collieries of the Inner Foothills Belt.**

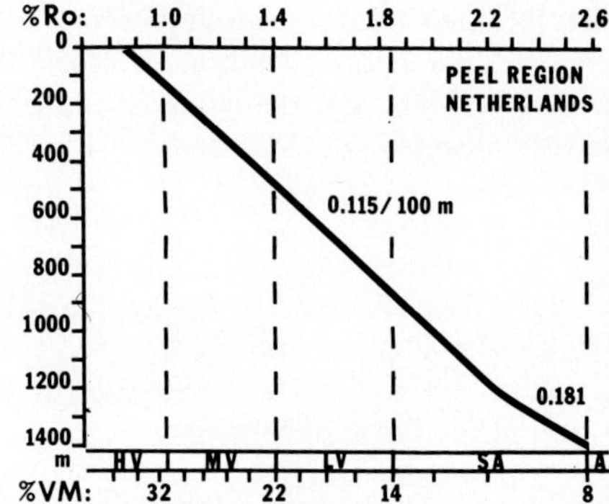


**Figure 5. The coalification curve of the Peel region in the Netherlands.**

by Kuyl and Patijn, a curve and not a straight line is obtained. This may indicate that the change in coalification is more directly related to a change in vitrinite reflectance than to a change in volatile matter.

The slope of the coalification curve indicates the rapidity of the change in rank with depth. A more shallow curve is related to a greater coalification gradient than a steeper curve, and this gradient can be expressed in terms of percent-reflectance (Ro) change per 100 m increase in depth. In the Peel curve, the coalification gradient is 0.115 per 100 m in the range from high-volatile coal to semianthracite, where it changes to 0.181 per 100 m.

When using the Peel curve as a reference for comparison, the coalification gradient of 0.115 per 100 m is employed to calculate the so-called *Peel rank (PR) ratio.* This ratio is obtained by dividing the coalification gradient of the curve under consideration by 0.115. Therefore, when the PR ratio is greater than 1, the rank changes faster than in the Dutch curve, and when it is less than 1, it changes at a slower rate.

## Coalification Curves of the Rocky Mountains and Inner Foothills Belt

The coalification curves of ten stratigraphic sections from different coal areas in the 600-mi-long Foothills Belt have been plotted in the diagrams of Figure 6. Their locations, which are shown by corresponding numbers on the map in the center, are as follows: (1) Peace River canyon; (2) Sukunka River, two boreholes; (3) Smoky River, No. 1 and No. 2 Mine areas; (4) Mountain Park; (5) Canmore coalfield; (6) Highwood River, Ford Collieries; (7) Fording River, Eagle Mountain; (8) Line Creek Ridge; (9) Natal Ridge; and (10) Sparwood Ridge.

All sections are Late Jurassic to Early Cretaceous in age and belong to the Kootenay Formation in the south (diagrams 5 to 10) and the younger Luscar Formation in the north (diagrams 1 to 4). The black dots in the diagrams refer to 94 individual coals of which the rank has been determined from the vitrinite reflectance (see Table 2). The dashed curves plotted through the dots show the relation between rank and stratigraphic position in each section, between the upper and lowermost coals present. The entire rank interval therefore is illustrated, and can be read in terms of percent reflectance on the abscissa at the top and in terms of percent volatile matter on the abscissa at the bottom of Figure 6. On the latter, the ASTM rank classes have been indicated also; these are HV, MV, and LV, for high-volatile, medium-volatile, and low-volatile bituminous coal with SA for semianthracite and A for anthracite.

Each diagram also shows the coalification curve of the Peel, which is marked with the solid line that has the same point of origin as the dashed curve. The coalification gradient and PR ratio of each curve are indicated in numerals, the

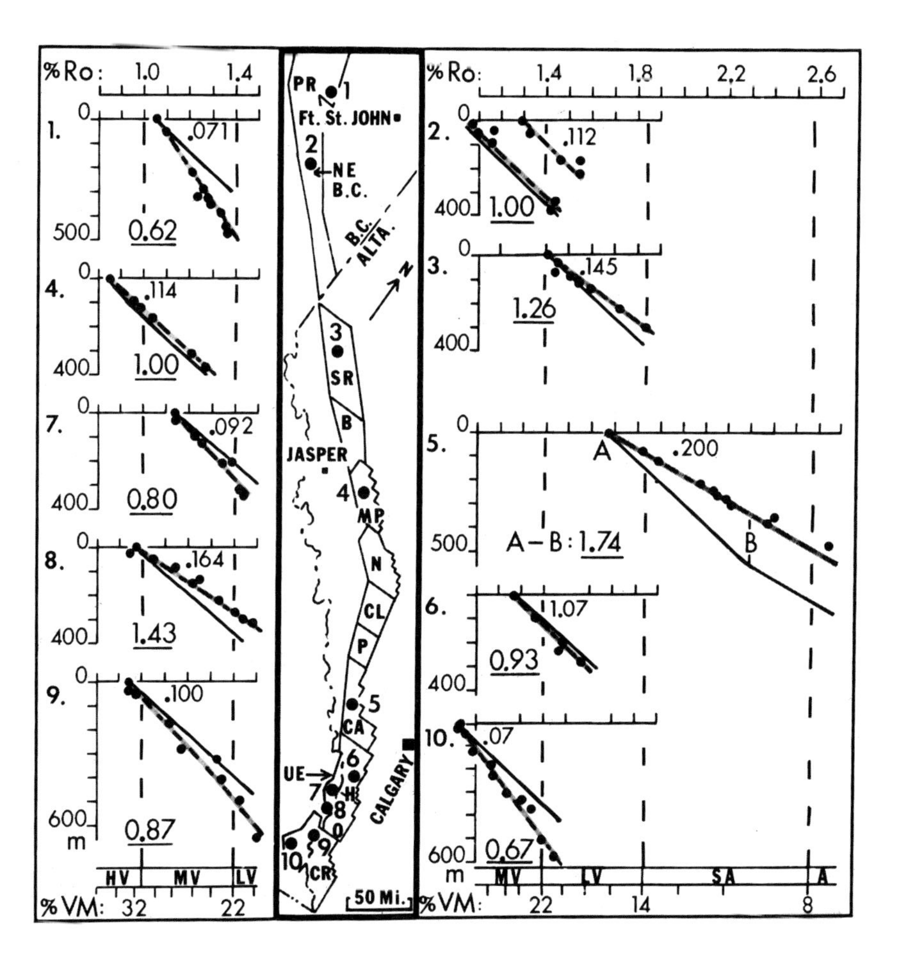


**Figure 6. Coal rank changes in 10 stratigraphic sections of the Rocky Mountains and Inner Foothills Belt (Peel reference curve is shown in solid line—0.62 indicates Peel rank ratio, 0.71 is coalification gradient in terms of percent Ro per 100 m; PR, SR, and so forth, are provincial coal areas according to MacKay, 1947. For locations of sections, see Table 2 and text).**

TABLE 2. RECORD OF COAL SAMPLES USED IN STRATIGRAPHIC RANK STUDY

| Sample no. | Name of section and location of coal seam | Stratigraphic position (ft) | (m) | Percent $Ro^{max}$ |
|---|---|---|---|---|
| | 1. Peace River canyon area; Gething Formation (Stott, 1969) | | | |
| PR-12 | Upper Coal—Moosebar Creek | 0 | 0 | 1.05 |
| PR-13 | Coaly Layer—Moosebar Creek | 170 | 51 | 1.09 |
| PR-11 | Upper Coal—Peace River Canyon, north shore at dam | 737 | 221 | 1.21 |
| PR-8 | Coal, 236 ft below PR-11 | 937 | 292 | 1.26 |
| PR-7 | Coal, 329 ft below PR-11 | 1,066 | 319 | 1.23 |
| PR-6 | Grant seam (?), 373 ft below PR-11 | 1,110 | 333 | 1.27 |
| PR-5 | Coal, 400 ft below PR-11 | 1,137 | 341 | 1.28 |
| PR-2 | Upper Twin, east bank of canyon | 1,280 | 384 | 1.33 |
| PR-3 | Lower Twin, east bank of canyon | 1,294 | 388 | 1.33 |
| PR-17 | Coal, bottom of canyon at foot of dam | 1,490 | 447 | 1.36 |
| PR-16 | Coal, 70 ft below PR-17 (Murray Horizon?) | 1,560 | 468 | 1.36 |
| | 2. Sukunka River area, Boreholes S 44 and D1; Commotion and Gething Formations | | | |
| Suk-8 | Coal cut at 232 ft in S 44 | 0 | 0 | 0.99 |
| Suk-9 | Coal cut at 251 ft in S 44 | 19 | 6 | 1.08 |
| Suk-10 | Coal cut at 304 ft in S 44 | 72 | 22 | 1.09 |
| Suk-11 | Coal cut at 385 ft in S 44 | 153 | 46 | 1.17 |
| Suk-12 | Coal cut at 411 ft in S 44 | 179 | 54 | 1.10 |
| Suk-13 | Coal cut at 541 ft in S 44 | 309 | 93 | 1.16 |
| Suk-14 | Bird seam cut at 1,347 ft in S 44 | 1,115 | 335 | 1.44 |
| Suk-15 | Chamberlain seam cut at 1,504 ft in S 44 | 1,272 | 381 | 1.42 |
| Suk-5 | Coal cut at 72 ft in D1 | 0 | 0 | 1.29 |
| Suk-4 | Coal cut at 293 ft in D1 | 160 | 48 | 1.32 |
| Suk-3 | Coal cut at 800 ft in D1 | 560 | 168 | 1.55(?) |
| Suk-2 | Coal cut at 815 ft in D1 | 570 | 171 | 1.46 |
| Suk-1 | Coal cut at 1,012 ft in D1 | 725 | 218 | 1.55 |
| | 3. Smoky River area, No. 1 and No. 2 Mine districts; Luscar Formation | | | |
| WC-32-I | No. 11 seam; No. 2 Mine district | 0 | 0 | 1.40 |
| WC-31-VI | No. 10 seam; No. 2 Mine district | 110 | 33 | 1.45 |
| Sky-6 | No. 7 seam; No. 1 Mine district | 250 | 75 | 1.43 |
| Sky-5 | No. 6 seam; No. 1 Mine district | 255 | 77 | 1.43 |
| Sky-4 | No. 5 seam; No. 1 Mine district | 300 | 90 | 1.50 |
| Sky-3 | No. 4 seam; No. 1 Mine district | 380 | 114 | 1.54 |
| Sky-2 | No. 3 seam; No. 1 Mine district | 460 | 138 | 1.59 |
| Sky-G | No. 2 seam; No. 1 Mine district | 740 | 222 | 1.72 |
| Sky-I | No. 1 seam; No. 1 Mine district | 980 | 294 | 1.83 |
| | 4. Mountain Park area; Luscar Formation (MacKay, 1929, 1930) | | | |
| MtP-1 | 2 Foot seam | 0 | 0 | 0.85 |
| MtP-4 | Lee seam | 324 | 97 | 0.96 |
| MtP-5 | Kennedy seam | 450 | 135 | 0.98 |
| MtP-3 | Seam overlying No. 1(?) | 466 | 140 | 0.98 |
| MtP-2 | No. 1(?) seam | 553 | 166 | 1.04 |
| MtP-7 | No. 4 seam | 1,040 | 312 | 1.21 |
| MtP-6 | No. 5 seam | 1,230 | 369 | 1.27 |
| | 5. Cascade area, Windridge section and Canmore coalfield; Kootenay Formation (Norris, 1957, 1971) | | | |
| CAN-1 | Marker seam | 0 | 0 | 1.67 |
| CAN-2 | Stewart seam | 242 | 73 | 1.82 |
| CAN-11 | Granger seam | 408 | 122 | 1.89 |
| CAN-9 | Upper Marsh seam | 717 | 215 | 2.08 |
| CAN-5 | Lower Marsh seam | 793 | 238 | 2.14 |
| CAN-6 | Seamlet in Tunnel 16, below CAN-5 | 848 | 254 | 2.15 |
| CAN-7 | Seamlet in Tunnel 16, below CAN-6 | 933 | 280 | 2.19 |
| CAN-3 | Wilson seam | 975 | 293 | 2.23 |
| CAN-8 | Big seam in Tunnel 16 | 975 | 293 | 2.22 |
| CAN-10 | Cairnes seam | 1,182 | 355 | 2.41 |
| CAN-14 | No. 6 seam | 1,268 | 380 | 2.38 |
| CAN-12 | Wind Creek seam | 1,591 | 477 | 2.65 |

TABLE 2 (CONTINUED)

| Sample no. | Name of section and location of coal seam | Stratigraphic position (ft) | (m) | Percent $Ro^{max}$ |
|---|---|---|---|---|
| | 6. Highwood River area, Ford Collieries section; Kootenay Formation (Ro from fuel ratios of Table 2 *in* Allan and Carr, 1947) | | | |
| 5 | 5 ft 1 in. seam of cut No. 2 | 0 | 0 | 1.28 |
| 4 | Glover seam, cut No. 4 | 300 | 90 | 1.37 |
| 6 | Connors seam, cut No. 1 | 650 | 195 | 1.49 |
| 7–8 | Douglas seam, cut No. 1 | 770 | 231 | 1.47 |
| 9–10 | Holt seam, cut No. 1 | 940 | 282 | 1.58 |
| | 7. Upper Elk Valley, Fording River section at Eagle Mountain; Kootenay Formation | | | |
| FR-2 | Seam No. 12 | 0 | 0 | 1.13 |
| FR-3 | Seam No. 11 | 106 | 32 | 1.14 |
| FR-4 | Seam No. 9 | 287 | 87 | 1.23 |
| FR-6 | Seam No. 7 | 406 | 123 | 1.25 |
| FR-5 | Seam No. 5 | 670 | 201 | 1.38 |
| FR-7 | Seamlet below No. 5 | 700 | 212 | 1.34 |
| FR-8 | Seam No. 2 | 1,110 | 333 | 1.42 |
| FR-9 | Seam No. 1 | 1,124 | 337 | 1.43 |
| | 8. Upper Elk Valley, Line Creek Ridge section; Kootenay Formation | | | |
| LC-9 | Second Coal above No. 2 seam | 0 | 0 | 0.97 |
| LC-8 | First Coal above No. 2 seam | 70 | 21 | 0.94 |
| LC-6 | No. 2 seam | 180 | 54 | 1.04 |
| LC-7 | No. 4 seam | 260 | 78 | 1.14 |
| LC-4 | No. 6 seam | 445 | 134 | 1.24 |
| LC-5 | No. 7 seam | 510 | 153 | 1.22 |
| CK-153 | No. 8 seam | 744 | 223 | 1.33 |
| LC-3 | No. 9 seam | 915 | 275 | 1.40 |
| LC-2 | No. 10 B seam | 990 | 297 | 1.44 |
| LC-1 | No. 10 A seam | 1,026 | 308 | 1.49 |
| | 9. Crowsnest area, Natal Ridge section; Kootenay Formation | | | |
| Mi-1 | D seam | 0 | 0 | 0.94 |
| Mi-3 | Upper C seam | 130 | 39 | 0.94 |
| Mi-2 | Lower C seam | 150 | 45 | 0.97 |
| Mi-4 | A seam | 567 | 170 | 1.12 |
| Mi-6 | Upper 5 seam | 930 | 279 | 1.17 |
| Mi-9 | No. 6 seam | 1,048(?) | 314 | 1.33 |
| Mi-7A | No. 7 seam | 1,360 | 408 | 1.35 |
| Mi-8 | No. 8 seam | 1,622 | 487 | 1.43 |
| Br-S | No. 10 seam | 2,130 | 639 | 1.50 |
| | 10. Crowsnest area, Sparwood Ridge section; Kootenay Formation | | | |
| BSW-4 | D seam | 0 | 0 | 1.04(?) |
| CQ-180 | Upper C seam | 74 | 22 | 1.03 |
| CQ-179 | Lower C seam | 136 | 41 | 1.05 |
| BSW-62 | B seam | 409 | 123 | 1.09 |
| CQ-177 | A seam | 615 | 184 | 1.17 |
| CSW-193 | No. 1 seam | 750 | 225 | 1.18 |
| CSW-165 | No. 2 seam | 761 | 228 | 1.18 |
| CQ-163 | No. 5 seam | 997 | 299 | 1.24 |
| CSW-111 | No. 6 seam | 1,081 | 324 | 1.31 |
| CQ-178 | No. 7 seam | 1,201(?) | 360 | 1.35 |
| CQ-162 | No. 9 seam | 1,669 | 501 | 1.40 |
| CQ-161 | No. 10 seam | 1,910 | 573 | 1.45 |

first in the upper right and the second (underlined) in the lower left of the diagrams. With a PR ratio of 1.00, the two curves coincide; when less than 1.00, the dashed curve lies below the Peel curve, and when more than 1.00, it lies above it.

From the ten diagrams of Figure 6 the following observations can be made:

1. In all sections, there exists a linear relation between rank and stratigraphic position, when the rank is expressed in percent vitrinite reflectance. This is in conformity with the Peel curve, but the change in slope at about 10 percent volatile matter of that curve (Fig. 4) is probably not present here. Diagram 5 of the Canmore coalfield shows no apparent change in slope, possibly as far down as 8 percent volatile matter. A similar diagram published by Norris (1971) also reveals no change, even though the fuel ratios on the whole coal have been used as rank parameters.

2. The slope of the dashed curve varies between the different diagrams. This variation is expressed in different PR ratios, which range from a low of 0.62 in diagram 1 to a high of 1.74 in diagram 5. This means that in diagram 1 the coalification gradient is a little more than half that of the Peel, and in diagram 5 it is greater than 1-1/2 times the Peel. The relative position of the two curves in the diagrams clearly illustrates this.

3. A coalification curve with a steep slope is, in general, more favorable for a greater presence of medium-volatile coking coals than a curve with a shallow slope. A comparison shows that the medium-volatile coals at Line Creek (No. 8) occur in a stratigraphic interval of only 250 m, while those at Sparwood (No. 10) are present in 500 m. As a result, there is less medium-volatile coal at Line Creek than at Sparwood. There are exceptions to this observation, notably in the section of diagram 9, but there is little doubt that in each area the coalification gradient plays an important role in the availability of coking coals of certain rank.

### Coalification in Relation to Temperature Gradient and Thermal Conductivity

Studies in coal metamorphism carried out during the past twenty years in various parts of the world have clearly demonstrated that in bituminous coals the increase in rank is the result of increasing rock temperature rather than pressure caused by overburden or folding (Hacquebard and Donaldson, 1970). Heat appears to be the prime agent involved in the coalification process, and from this point of view, Hilt's law can be explained readily by the normal rise in temperature with increasing depth of burial. However, in order to obtain different coalification gradients, variations in temperature gradient and heat flow are required. These variations occur in the coal measures because of changes in the geothermal gradient and because of substantial differences in the thermal conductivity of clastic rocks (Ammosov, 1970; Damberger, 1968).

Correlations between the coalification and geothermal gradients have been demonstrated by Kuyl and Patijn (1961), Teichmüller and Teichmüller (1966, 1967), and by Damberger (1968). The former calculated the geothermal gradient of the Peel region by comparing its coalification gradient with that of a coal sequence with known temperature gradient. The calculated and measured gradients were remarkably close, namely 5.0°C per 100 m as compared to 4.2°C per 100 m.

Damberger (1968), in detailed studies of the coal measures of the Saar Basin, showed that thick sandstone and conglomerate beds are associated with small coalification gradients because of the higher thermal conductivity of these rocks as compared to shale and coal. A higher thermal conductivity causes a lower temperature gradient. Therefore, in limited areas where a constant terrestrial heat flow may be assumed (that is, in boreholes), coals with sandstone predominant as intervening strata show less difference in rank than coals that are separated by shale and other carbonaceous layers.

From the preceding discussion, it would appear that the differences in the coalification gradients observed in the Rocky Mountains and Foothills Belt are primarily related to variations in the geothermal gradient. Deviations in the rank of individual seams from the projected linear curve, on the other hand, may be due to the presence of thick sandstone members in the particular coal-bearing sequence.

In Figure 6, only sections 3, 5, and 8 possess a PR ratio that is larger than 1, signifying a larger coalification gradient than that present in the Peel region. Their geothermal gradients are therefore considered larger than 4.2°C per 100 m. It is interesting to note that at Smoky River (No. 3) and Canmore (No. 5) this higher gradient not only caused a more rapid increase in coalification, but also a rank range that is of a higher order, namely the range of low-volatile bituminous coal and semianthracite.

As regards the Canmore area, Norris (1971) also considers greater thermal metamorphism as the prime cause for the presence of semianthracites in this region. This greater thermal effect can be related only to a higher geothermal gradient and not to a higher temperature caused by greater initial depth of overburden. From the palinspastic maps with isopachs of the Blairmore and Kootenay Formations constructed by Norris (1964, 1971), it can be seen that no additional thickness of strata occurred at Canmore as compared to that postulated for sections 7 to 10, which encompass a lower rank increment. A higher geothermal gradient also is supported by the existence of a metamorphic halo outlined by isograds of staurolite-kyanite, garnet, and biotite that Price and Mountjoy (1970) have reported in the Rocky Mountains immediately west of the areas containing the highest rank coals (Norris, 1971).

More difficult to explain is the unexpectedly high PR ratio of 1.43 and the corresponding large geothermal gradient at Line Creek (No. 8). The sections at Fording River (No. 7), Sparwood Ridge (No. 10), and Natal Ridge (No. 9), situated to the north and south of section 8 show comparable PR ratios that lie between 0.87 and 0.67. The Line Creek section, moreover, generally has a greater percentage of sandstone than is present in any of the other sections mentioned. More sandstone tends to retard the coalification gradient when the same amount of heat is available. Here, a greater-than-average coalification gradient is present, and only a considerably increased geothermal gradient can account for this, at least to the authors' knowledge.

In the other sections of Figure 6, the PR ratio is equal to or less than that of the Peel region. Accordingly, the geothermal gradients at the time of coalification have been 4.2°C per 100 m or less. Throughout the Rocky Mountain and Foothills region, therefore, there have been marked variations in the geothermal gradient. The lowest gradient probably occurred in the area of the Peace River canyon (with a PR ratio of 0.62), and the highest at Canmore (with a ratio of 1.74).

## SEAM CORRELATIONS BASED ON RANK OBTAINED FROM VITRINITE REFLECTANCE MEASUREMENTS

When the coalification curve is shallow, the Ro rank can be utilized for the correlation of seams, because then the reflectance changes rapidly over relatively short intervals. This is the case at Canmore, where a shallow curve with the highest PR ratio of the entire Foothills region has been encountered, specifically, a ratio of 1.74 (see diagram 5 in Fig. 6).

The Canmore coalfield lies in an intensely folded and faulted area, and the interpretation of the geological structure, which is of great importance to the mining industry, has been a difficult task. Following the original studies made by MacKay in 1934, a wealth of new information has become available from continued underground and strip mining. From this information and extensive personal observations, Norris (1957, 1971) has compiled a new geological map with structural interpretations of the Cascade coal basin. A much-simplified version of this map is reproduced in Figure 7. It shows the outcrops and fold patterns of seven major seams, which in descending stratigraphic order are the Marker (240), Stewart (160), Sedlock-Morris-Granger (300), Upper Marsh-Carey (70), Lower Marsh-No. 4 (175), Wilson (50), and Big seams. The number in brackets in this sequence refers to the average thickness in feet of the seam intervals, as recorded by Norris (1971).

The vitrinite reflectance values on 27 samples collected from different localities are given in Table 3. They vary from 1.65 to 2.26 percent and show for each seam specific ranges that do not overlap. These ranges are indicated in the upper left-hand corner of Figure 7. They were obtained from different samples that were known to represent the same seam at different locations. These samples are marked with the solid dots, while the open circles refer to locations where the identification of the seam was not definitely known, but is here suggested on the basis of the Ro value. For example, in the Stewart seam, the range in Ro of 1.75 to 1.82 percent was obtained on nine samples from three known localities, namely at 10, 19, and 20. The unknown coal at 23 is correlated with the Stewart because it has a reflectance of 1.81 percent.

The other correlations have been obtained in a similar manner. They include the identification of the Sedlock and Marker seams in the Upper Canmore Creek area (locations 24 through 27) and the No. 4 seam in old prospect pits situated northwest of the Wilson Mine (locations 12 and 13). Also proposed is the correlation of the Big Seam (at 4) with the Wilson seam (at 11), and the Carey with either the Upper Marsh or the Lower Marsh, or with both seams combined. The stratigraphic distance of these seams is only 70 ft, which is insufficient for a Ro rank separation. Since lateral variations in reflectance of up to 0.09 percent occur within individual seams, there exists a minimum distance below which the Ro range of one seam overlaps onto the other. This distance is related to the slope of the coalification

TABLE 3. RECORD OF COAL SAMPLES USED IN STUDY OF SEAM CORRELATION BY RANK IN CANMORE COALFIELD

| Index no. on Figure 7 | Sample no. | Obtained from | Name and location of seam | Percent $Ro^{max}$ |
|---|---|---|---|---|
| 1 | CAN-9 | Underground mine | Upper Marsh; No. 3 Mine | 2.09 |
| 2 | WC-XIII | Underground mine | Upper Marsh; No. 3 Mine | 2.07 |
| 3 | CAN-5 | Underground mine | Lower Marsh; Tunnel 16 No. 3 Mine | 2.14 |
| 4 | CAN-8 | Underground mine | Big seam; Tunnel 16 No. 3 Mine | 2.22 |
| 5 | CAN-11A | Trench | Granger; S. side Stewart Creek | 1.94 |
| 6 | CAN-11 | Prospect | Granger; S. side Fall Creek | 1.89 |
| 7 | CAN-4 | Strip mine | No. 4; Dry Lake area | 2.18 |
| 8 | CAN-26 | Strip mine | No. 4; Dry Lake area | 2.19 |
| 9 | CAN-43 | Old adit | Morris; 2,800 ft S. of Wilson Slope | 1.99 |
| 10 | CAN-41 | Old adit | Stewart; 1,700 ft SW of Wilson Slope | 1.82 |
| 11 | CAN-27 | Underground | Wilson; Wilson Mine | 2.25 |
| 12 | CAN-49 | Prospect | Carey(?); 2,400 ft NW of No. 4 Slope at 17,370 N; 15,370 E. | 2.18 |
| 13 | CAN-50 | Prospect | Carey(?); S. bank Bow River at 18,400 N.; 14,680 E. | 2.14 |
| 14 | CAN-40 | Prospect | Sedlock; 1,300 ft W. of CAN-49 at 17,300 N.; 14,100 E. | 1.94 |
| 15 | CAN-45 | Old adit | Carey; 1,200 ft W. of CAN-50 at 18,600 N.; 13,500 E. | 2.11 |
| 16 | CAN-44 | Prospect | Sedlock; No. 2 Mine site at 20,400 N.; 10,000 E. | 1.97 |
| 17 | WC-XIV | Underground mine | No. 4; No. 2 Mine | 2.11 |
| 18 | CAN-1 | Adit | Marker; 1,700 ft SE of Stewart Mine at 15,000 N.; 12,200 E. | 1.67 |
| 19 | CAN-2 | Underground mine | Stewart; No. 2 Stewart Mine Slope | 1.82 |
| 20 | CAN-53 | Adit | Stewart; 200 ft NE of Stewart Mine at 16,450 N.; 11,250 E. | 1.80 |
| 21 | CAN-38 | Prospect | Marker(?); 2,400 ft NW of Stewart Mine at 17,700 N.; 9,250 E. | 1.74 |
| 22 | CAN-39 | Prospect | Unknown seam, correlated with Marker; 2,400 ft SW of No. 2 Mine at 19,750 N.; 7,750 E. | 1.72 |
| 23 | CAN-42 | Prospect | Unknown seam, correlated with Stewart; 2,300 ft SW of No. 2 Mine at 20,150 N.; 7,600 E. | 1.81 |
| 24 | CAN-51 | Strip mine | Unknown seam, correlated with Sedlock; 1,500 ft SW of No. 1 Mine at 23,100 N.; 4,500 E. | 1.93 |
| 25 | CAN-22 | Trench | Unknown seam, correlated with Marker; 2,200 ft SW of No. 1 Mine at 23,050 N.; 3,700 E. | 1.67 |
| 26 | CAN-15 | Trench | Unknown seam, correlated with Marker, 2,100 ft WSW of No. 1 Mine at 23,700 N.; 3,400 E. | 1.65 |
| 27 | CAN-19 | Trench | Unknown seam, correlated with Marker, 2,500 ft W. of No. 1 Mine at 24,500 N.; 2,950 E. | 1.71 |

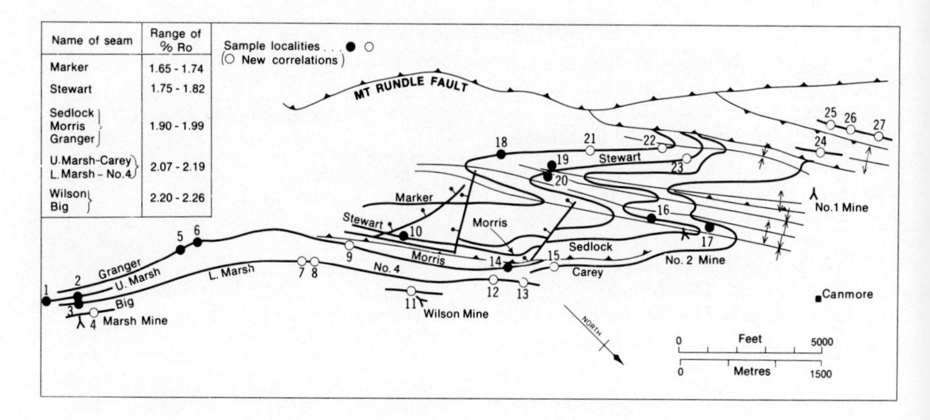


**Figure 7. Outline map of Canmore coalfield (after Norris, 1957, 1971), with vitrinite reflectance data of the major coal seams.**

curve. In the Canmore area, it is 120 ft, and seams that lie closer than this distance cannot be identified separately by this method.

At Canmore the coal operators have long been aware of the relation between rank and stratigraphic position and have successfully employed rank determinations to solve structural problems. Proximate analyses are used, and the introduction of vitrinite reflectance as a rank parameter is essentially a more refined and improved version of the existing method. It requires fewer samples, and high-ash as well as weathered coal can be used.

## ACKNOWLEDGMENTS

The authors have received much cooperation from the geological and engineering departments of the mining companies that are presently active in the Rocky Mountain region. Assistance was rendered with the collection of samples, and the stratigraphic information essential to this investigation was readily provided. The following companies are gratefully acknowledged: Brameda Resources-Coalition Mining Ltd., The Canmore Mines Ltd., Cardinal River Coals Ltd., Coleman Collieries Ltd., Crowsnest Industries Ltd., Fording Coal Ltd., Kaiser Resources Ltd., and McIntyre Coal Mines Ltd.

We also thank D. K. Norris of the Geological Survey of Canada, who greatly assisted our seam correlation project at Canmore by collecting critical samples, advising on new sample localities, and providing stratigraphic and structural information without which this study could not have been undertaken. He also reviewed the manuscript and offered helpful critical comments.

## REFERENCES CITED

Allan, J. A., and Carr, J. L., 1947, Geology of Highwood-Elbow area: Research Council Alberta Rept., no. 49, 74 p.

Ammosov, I. I., 1970, Stages of lithification of sedimentary rocks: Sheffield, England, Compte Rendu 6th Carboniferous Cong., 1967, v. 2, p. 403-415.

Damberger, H., 1968, Ein Nachweis der Abhängigkeit der Inkohlung von der Temperatur: Brennstoff-Chemie 3, Bd. 49, p. 73-77.

Hacquebard, P. A., and Donaldson, J. R., 1970, Coal metamorphism and hydrocarbon potential in the Upper Paleozoic of the Atlantic provinces, Canada: Canadian Jour. Earth Sci. 7, p. 1139-1163.

Hilt, C., 1873, Die Beziehungen zwischen der Zusammensetzung und den technischen Eigenschaften der Steinkohlen: Zeitschr. Ver. Deutscher Ingen., Bd. 17, Ht. 4, p. 194-202.

Jansa, L. F., 1971, Depositional history of the coal-bearing Upper Jurassic-Lower Cretaceous Kootenay Formation, southern Rocky Mountains, Canada, *in* Geol. America Guidebook, Field Trip No. 1, Cascade and Crowsnest Coal Basins, May 10-12, 1971: p. 1-26.

Kötter, K., 1960, Die mikroskopische Reflexionsmessung mit dem Photomultiplier und ihre Anwendung auf die Kohlenuntersuchung: Brennstoff-Chemie 41, p. 263-272.

Kuyl, O. S., and Patijn, R. J. H., 1961, Coalification in relation to depth of burial and geothermic gradient: Heerlen, Netherlands, Compte Rendu 4th Carboniferous Cong., 1958, v. 2, p. 357-365.

Latour, B. A., and Chrismas, L. P., 1970, Preliminary estimate of measured coal resources including reassessment of indicated and inferred resources in western Canada: Canada Geol. Survey Paper 70-58, 14 p.

MacKay, B. R., 1929, Mountain Park sheet: Canada Geol. Survey Map 208A, scale 1 in.: 1 mi.

——1930, Stratigraphy and structure of bituminous coalfields in the vicinity of Jasper Park, Alberta: Canadian Inst. Mining and Metallurgy Trans., v. 33, p. 473-509.

——1933, Geology and coal deposits of Crowsnest Pass area, Alberta: Canada Geol. Survey Summ. Rept., 1932, pt. B., p. 21-68.

——1934, Canmore area: Canada Geol. Survey Maps 322A and 323A, scale 1 in.: 800 ft.

——1947, Coal reserves of Canada (reprint of Chap. 1 and Appendix A of Rept. of the Royal Comm. on Coal, 1946): Ottawa, Edmond Cloutier, King's Printer, 113 p.

——1949, Atlas, Coal areas of Alberta: Canada Geol. Survey, 50 maps, scale 1 in.: 4 mi.

Norris, D. K., 1957, Canmore, Alberta: Canada Geol. Survey Paper 57-4, 8 p.

——1959, Type section of the Kootenay Formation, Grassy Mountain, Alberta: Alberta Soc. Petroleum Geology Jour., v. 7, no. 10, p. 223-233.

——1964, The Lower Cretaceous of the southeastern Canadian Cordillera: Bull. Canadian Petroleum Geology, v. 12, Field Conf. Guidebook Issue, p. 512-535.

——1971, The geology and coal potential of the Cascade coal basin, *in* A guide to the geology of the eastern Cordillera along the Trans Canada Highway between Calgary and Revelstoke: Banff, Alberta Soc. Petroleum Geology, Canadian Expl. Frontiers Symposium, p. 25-39.

Price, R. A., and Mountjoy, E. W., 1970, Geologic structure of the Canadian Rocky Mountains between Bow and Athabasca Rivers—A progress report: Canada Geol. Assoc. Spec. Paper no. 6, p. 7-25.

Stott, D. F., 1969, The Gething Formation at Peace River Canyon, British Columbia: Canada Geol. Survey Paper 68-28, 30 p.

Teichmüller, M., 1967, Inkohlungsuntersuchungen an kohleführenden Sedimenten des Tertiärs aus dem Oberrhein-Graben nördlich Worms: Oberrhein. Geol. Abh. 16, p. 11-15.
Teichmüller, M., and Teichmüller, R., 1966, Geological causes of coalification: Washington, D.C., Coal Science-Advances in Chemistry Ser. 55, p. 133-153.

SYMPOSIUM HELD AT G.S.A. ANNUAL MEETING IN MILWAUKEE, NOVEMBER 1971
MANUSCRIPT RECEIVED BY THE SOCIETY MARCH 9, 1973

Printed in U.S.A.

Geological Society of America
Special Paper 153


# Vitrinite Reflectance as an Indicator of Coal Metamorphism for Cokemaking

R. R. THOMPSON
AND
L. G. BENEDICT

*Coal and Coke Section*
*Homer Research Laboratories*
*Bethlehem Steel Corporation*
*Bethlehem, Pennsylvania 18016*

## ABSTRACT

The measurement of vitrinite reflectance has been employed as the standard measurement of coal metamorphism for more than a decade in the application of coal microscopy to the cokemaking industry. At Bethlehem Steel, however, it was recently found that the reflectance measurement of metamorphic stage must be restricted to truly reactive vitrinite, excluding pseudovitrinite, before acceptably accurate relations can be established between coal petrographic composition and coking properties. Pseudovitrinite, which we no longer class with vitrinite, apparently has undergone an unusual geologic history, and its reflectance measurement can be misleading as an indicator of metamorphic stage.

Restriction of the measurement of metamorphic stage to reactive vitrinite not only improves our ability to work out cokemaking relations but also reveals discontinuities in the coal metamorphic series that must have accompanied fundamental changes in coal constitution. At least three such discontinuities occur within the coking coal range, and these can be documented with cokemaking data. These findings form the basis of a practical system for the prediction of coking properties from petrographic analyses.

## INTRODUCTION

Research in the past decade has demonstrated that the accurate determination of the metamorphic stage, or rank, of coal is probably the most important factor in evaluating coals for industrial processes. Of course, rank can be expressed by a number of available parameters, and the particular type of measurement to be used depends primarily on the application being considered.

With respect to the cokemaking industry, the microscopic measurement of the reflectance of the vitrinitic constituents of coal has been developed over the past ten years as the method for measuring the metamorphic stage of coal, and this development is well documented in the literature (see references). The mechanics of the coal-reflectance analysis have been refined to the point where the technique is now fairly well standardized, both nationally and internationally. This measurement, which actually refers to the mean maximum reflectance of coal particles in oil, has been accepted by ASTM (1972) as a tentative standard and by the International Committee for Coal Petrology for publication in their *Handbook.*

Petrographers usually agree on the mechanics of the reflectance measurement, but not on components of coal that should be measured to characterize the metamorphic stage of coal. The objectives of the coal petrographer in the cokemaking industry have been to classify accurately the organic components of coal into reactive and inert substances with respect to their behavior in the coking process and to characterize accurately the metamorphic stage of the reactive components. Experience has shown that if these two goals are accomplished, good relations can be established between coal petrographic analyses and coke quality.

This paper describes progress that has been made toward these objectives at Bethlehem Steel Corporation. The work shows that limitation of the reflectance measurement to a selected fraction of vitrinite reveals significant discontinuities, which are related to significant differences in coking properties, in the coal metamorphic series.

Although previous investigators were aware of the existence of discontinuities, little practical significance was at first attached to these phenomena. For example, Seyler (1952) suggested the possibility of many breaks in the metamorphic series on the basis of the distribution of vitrinite reflectance measurements, but these breaks were not well documented by independent data. Stach (1953) recognized a "coalification step," or discontinuity, in Ruhr coals at the approximate boundary between high-volatile and medium-volatile coals, where the durainic bands are reported to decrease significantly in volatile-matter content. Stach's coalification step also marks the point at which exinite begins to increase in reflectance to approach the reflectance of vitrinite in higher rank coals. Stach further noted that this reflectance increase in the exinite, which probably includes devolatilization, could also account for the reduction in volatile-matter content of the durainic bands, since these bands are known to contain large percentages of exinite.

Gin and others (1963) described discontinuities that most closely relate to those that are described here. These authors subdivided the coking coals into three types, largely on the basis of the strength of the resulting coke. The three groupings they obtained were, approximately: (a) the low-rank coking coals, including those below high-volatile A bituminous in rank, (b) the medium-rank coking coals, including the high-volatile A bituminous through most of the medium-volatile bituminous coals, and (c) the high-rank coking coals, including the uppermost range of the medium-volatile bituminous coals and the low-volatile bituminous coals. Gin and others (1963) also found that each of these coal types required separate weighting in mathematical calculations before accurate estimates could be made of the coke strength to be expected from a coal blend.

None of the discontinuities found by previous authors correspond exactly to those described and documented in this paper. The results of our test programs show that significant differences in coking pressure as well as the coke properties of strength and size are related to the discontinuities revealed by reflectance measurements that exclude pseudovitrinite.

During the early 1960s, the accepted practice for determining the metamorphic stage of the reactive components in coal was to measure the reflectance of vitrinite in coal. However, work at Bethlehem Steel in recent years has shown that vitrinite consists, in fact, of two populations of constituents; one is truly reactive in coking, and the other is partly or entirely inert in coking. These two populations, referred to respectively as reactive vitrinite and pseudovitrinite, can be recognized and quantitatively defined microscopically, as described in previous literature (Benedict and others, 1968a, 1968b).

The chief practical result of our earlier studies, based on the identification of a pseudovitrinite component, is that the coking properties of a given coal can be predicted from the metamorphic stage of that coal as determined by reflectance measurements restricted to vitrinite, the reactive component of coal, as opposed to the inert or partially inert pseudovitrinite.

As a direct consequence of these findings, the use of selective coal reflectance was refined and broadened as a practical tool for determining the rank of coals and for predicting their suitability for coking. This tool has since become a routine part of Bethlehem Steel's system of coal evaluation for coke plant operations.

## MEASURING REFLECTANCE IN RELATION TO THE PSEUDOVITRINITE-VITRINITE SUBDIVISION

Restriction of the reflectance measurement to vitrinite, the reactive component, presumes that a separation can be made between pseudovitrinite and vitrinite. Since the criteria for identifying these two populations of vitrinitic constituents have been dealt with in detail in the literature (Benedict and others, 1968a, 1968b), it will suffice to briefly review them here.

Pseudovitrinite is separated from vitrinite primarily on the basis of its higher reflectance in a given coal sample, as shown in Figure 1. When the brighter pseudovitrinite particles are essentially structureless, identification is sometimes aided by a shadowy or milky appearance that is suggestive of destroyed cellular structures, as shown in the bottom photomicrograph in Figure 1. The separation of pseudovitrinite and reactive vitrinite becomes more difficult, of course, as the reflectances of these constituents approach each other. However, as Figure 2 illustrates, the identification of pseudovitrinite is further aided by the fact that the brightest of these constituents tend to contain either remnant cellular structures or slitted structures, the latter of which are perhaps devolatilization cracks.

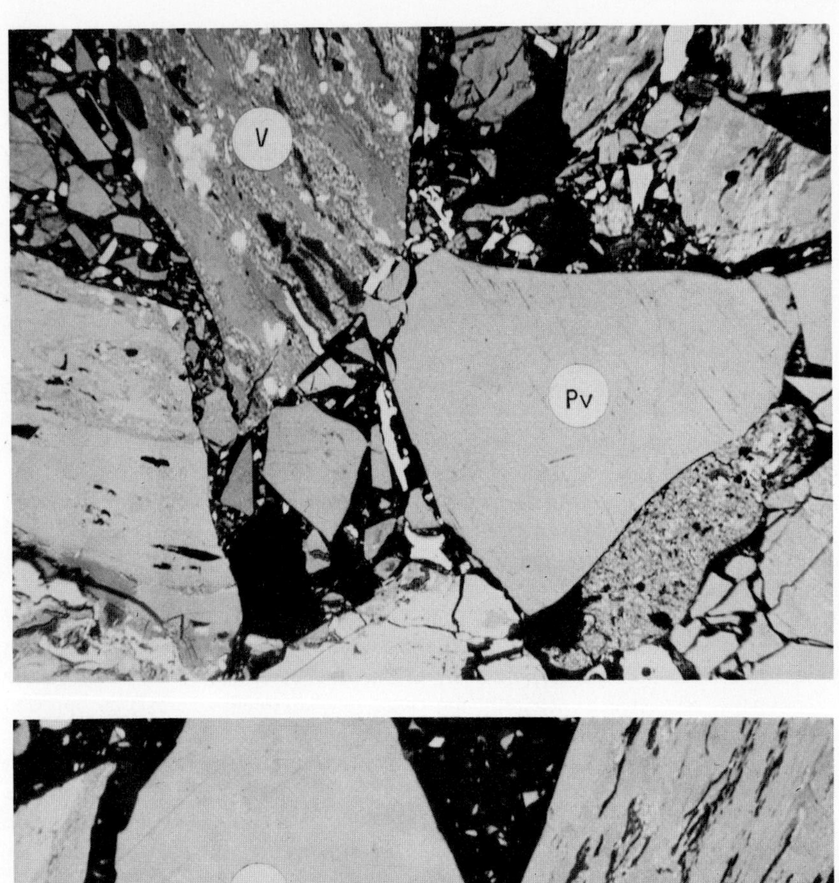


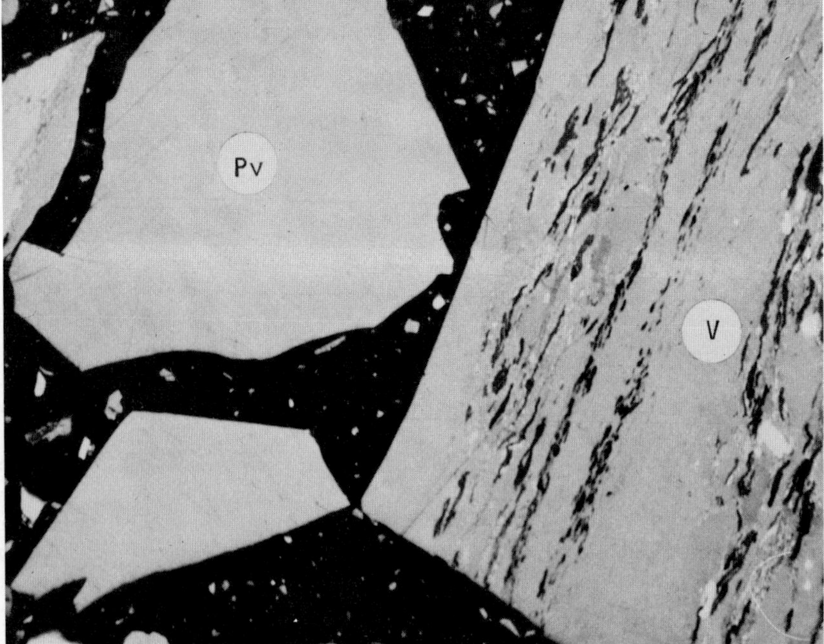


**Figure 1. The comparative brightness of reactive vitrinite and pseudovitrinite on polished coal surfaces (× 315). The pseudovitrinite (Pv) is brighter than the vitrinite (V) in both cases.**

Pseudovitrinite has still other secondary features which are helpful in identification. Thus, even when reflectances are close, the two populations can be separated and quantified by trained petrographers.

Pseudovitrinite is inherent in all coal seams that we have studied, and we consider it to be most likely a product of partial oxidation that occurred early in coal formation, perhaps in the peat swamp. Such an origin would account for its higher reflectance, higher oxygen content, and behavior as an inert or semi-inert substance in the coking process.

The different roles played by pseudovitrinite and vitrinite in the coking process can be illustrated by examining the structure and properties of cokes made from these two groups as well as by comparing the coking characteristics, such as pressure, of two coals containing significantly different percentages of these constituents.

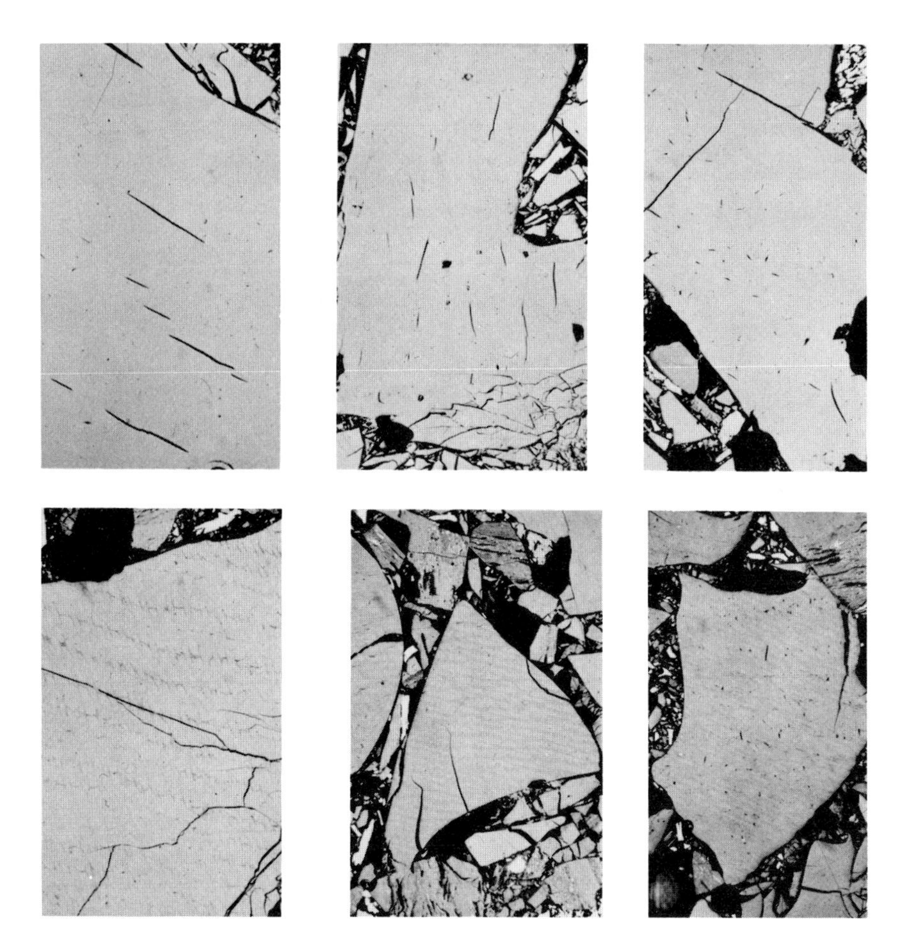

**Figure 2. Typical microstructures in pseudovitrinite (× 315); remnant cellular structures (below) and slitted structures (above).**

The appearance of vitrinite and pseudovitrinite in a coke are illustrated in Figure 3. Figure 3A shows a particle of pseudovitrinite that has remained almost completely inert during the coking process. It has retained its original morphology, and its boundaries contact sharply with the surrounding coke mass, indicating that little melting has taken place. The surrounding anisotropic, granular-appearing coke walls are typical products of vitrinite that has melted and resolidified in the coking process. Figures 3A, B, and C, which are pictures of the same coke, illustrate

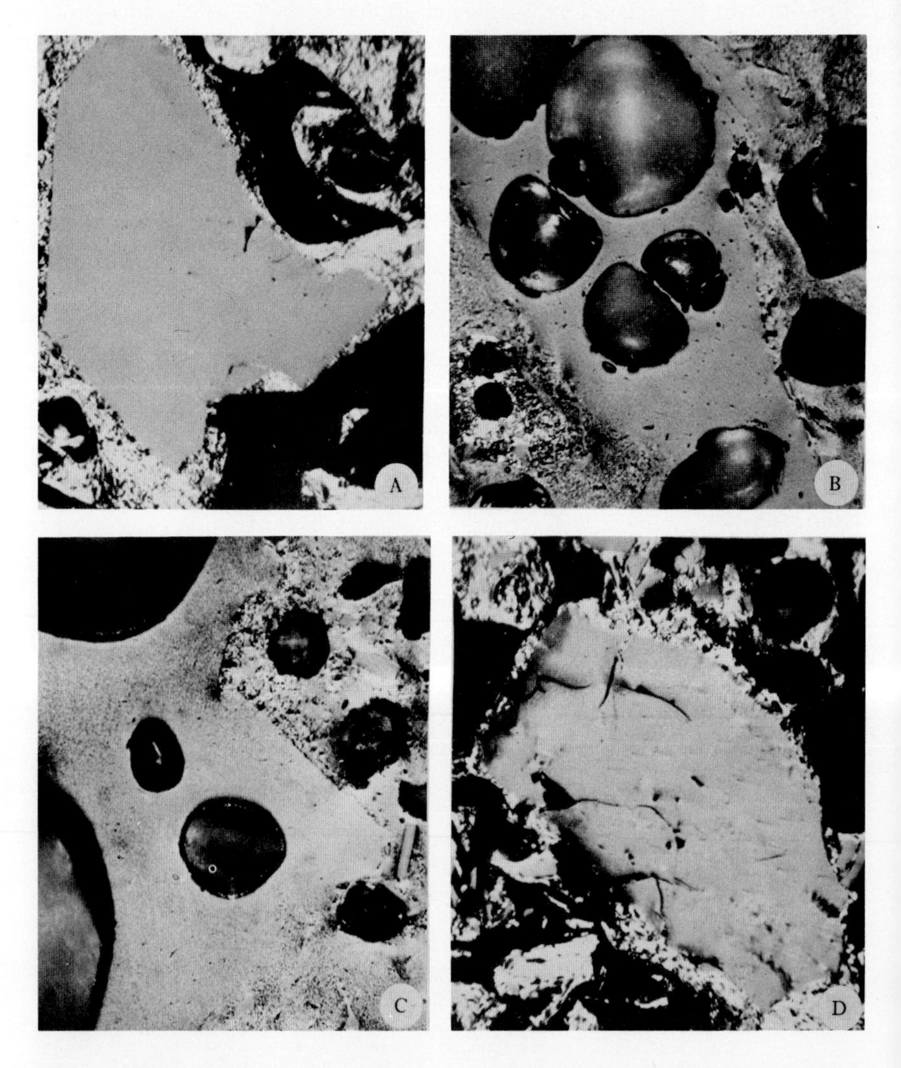


**Figure 3. High-temperature coke products from pseudovitrinites and weathered vitrinite (× 236). (A) Coked product from highly altered pseudovitrinite; (B) coked product from moderately altered pseudovitrinite; (C) coked product from slightly altered pseudovitrinite; (D) coked product from weathered vitrinite.**

the range in inertness of pseudovitrinite. In Figure 3B, pores that have developed in the pseudovitrinite suggest some degree of reaction; Figure 3C shows a pseudovitrinite particle that has reacted to a greater extent and has developed some degree of anisotropy. For comparison, Figure 3D shows a particle of weathered coal as it appears in coke. This particle exhibits a partially weathered halo that appears dark in this picture, as well as the characteristic deep fractures and high relief that distinguish weathered particles from pseudovitrinite particles.

Table 1 illustrates the importance of the differences in reactiveness between vitrinite and pseudovitrinite in the development of a system for predicting coke quality from microscopic analyses. On the one hand, when the prediction of coke stability was based on the reflectance of total vitrinite, using the method of Schapiro and others (1961), the prediction in the case of Coal B was quite divergent from the actual test-oven result, that is 38 versus 20. On the other hand, when the selective vitrinite reflectance method is used, that is, when the largely inert pseudovitrinite was excluded from the reflectance measurement and relegated to the inert category, the coke-stability prediction for Coal B was in close agreement with the test-oven result. The significant lowering of the reflectance of Coal B when pseudovitrinite was excluded from the reflectance measurement is attributed to the fact that the reflectance of the pseudovitrinite constituent averaged more than 0.1 percent higher than the reflectance of the true, or reactive, vitrinite constituent of this coal. In contrast, when pseudovitrinite was a much less significant component, as in Coal A, the method employed by Schapiro and others was able to provide a prediction that was in agreement with the test-oven result.

Application of the Bethlehem Steel system of coal petrographic analysis that developed out of the kind of results exemplified in Table 1 led to a substantial improvement in the prediction of coke quality.

The reasons for the greater effectiveness of this system of analysis are (a) the reflectance analysis is restricted to reactive vitrinite, (b) the inert proportion of the pseudovitrinite is determined according to procedures described by Benedict

TABLE 1. COMPARISON OF PREDICTIONS OF COKE STRENGTH USING TWO DIFFERENT METHODS OF COAL PETROGRAPHIC ANALYSIS

| | Conventional method | | | |
|---|---|---|---|---|
| | Reflectance of total vitrinite (%) | Organic inert content (%) | Predicted* coke stability (%) | Measured stability of test oven coke (%) |
| Coal A | 0.88 | 12.2 | 31 | 31 |
| Coal B | 0.88 | 18.6 | 38 | 20 |
| | Bethlehem Steel's method† | | | |
| | Reflectance of reactive vitrinite (%) | Effective inert content (%) | Predicted coke stability (%) | Measured stability of test oven coke (%) |
| Coal A | 0.86 | 20.5 | 31 | 31 |
| Coal B | 0.81 | 37.1 | 18 | 20 |

*Predicted using method of Shapiro and others (1961).
† Benedict and others (1968a).

and others (1968a), and (c) new and much simplified relations between coke stability and coal petrographic analyses are employed (Benedict and others, 1968b).

Finally, along with providing the strictly practical result of a more effective system of coal analysis for coking, the restricting of reflectance measurement to reactive vitrinite is, as a realistic gage of the degree of coalification, a helpful concept in more broadly geologic terms as well.

## DISCONTINUITIES IN THE COAL METAMORPHIC SERIES

If the reflectance measurement is restricted to reactive vitrinite and the measurements of these constituents are plotted against an independent measurement of coal rank, surprising relations emerge, as shown in Figure 4. The independent rank measurement in this figure is the volatile-matter content of the coal, calculated to a dry, ash-free basis. The lines are drawn by eye to best fit the points representing analyses of about 125 coal samples. Each point represents an average of 25 or more reflectance measurements on individual reactive vitrinite particles in each sample. The large spread in points around the line is probably due primarily to the reproducibility of the volatile-matter measurement rather than to the reproducibility of the more accurate reflectance measurement.

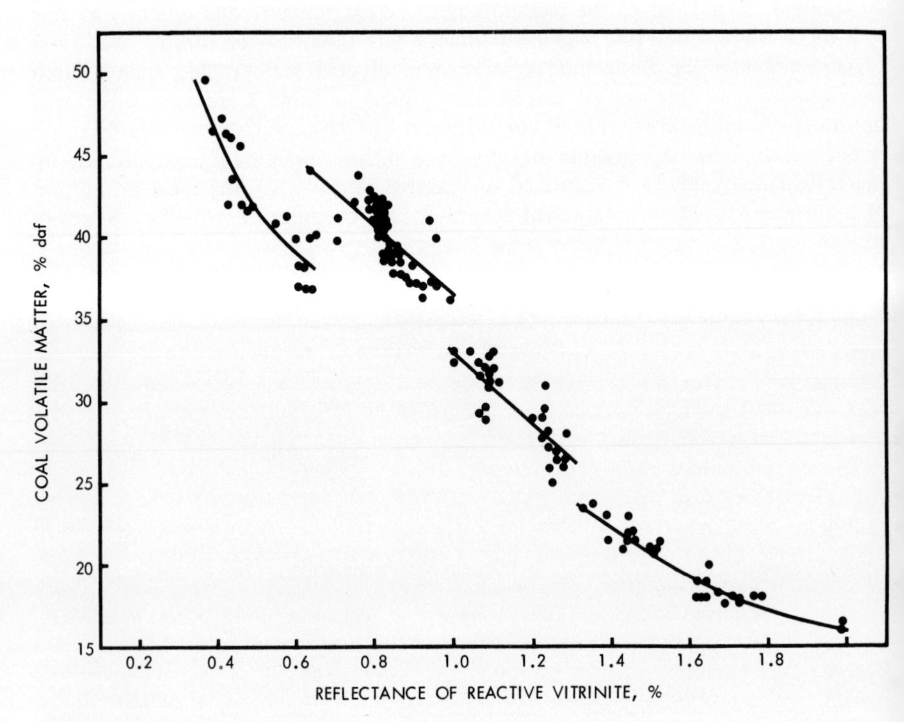


**Figure 4. Discontinous relation between reflectance of reactive vitrinite and coal volatile-matter content.**

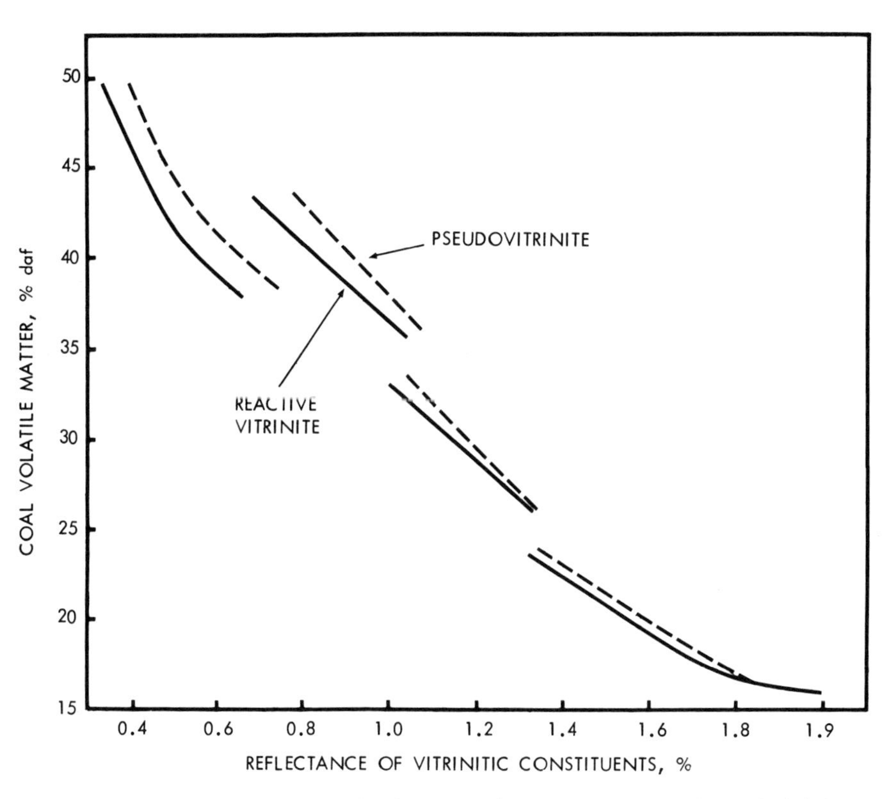


**Figure 5. Discontinuous relations of reflectances of reactive vitrinite and pseudovitrinite to coal volatile-matter content.**

Figure 4 shows that the relation between reactive-vitrinite reflectance and coal volatile-matter content is represented best by a series of discontinuous lines rather than by a single curve. These lines are separated by the distinct discontinuities that occur around the 0.7, 1.0, and 1.35 percent reflectance levels. The practical significance of these discontinuities is that they occur respectively at the approximate boundaries between the following coals as classified by ASTM standards:(a) high-volatile A and high-volatile B coals, (b) high-volatile and medium-volatile coals, and (c) medium-volatile and low-volatile coals.

The discontinuities shown in Figure 4 are also evident in the relation between the reflectance of pseudovitrinite and the coal volatile-matter content, as shown in Figure 5. It can also be seen here that the reflectance of pseudovitrinite is always higher than that of reactive vitrinite and that the difference in the reflectance of these two populations of constituents diminishes with an increase in coal rank. In the low-volatile coals, the two reflectances are almost identical. In general, pseudovitrinite is increasingly inert in coking as its reflectance increases with respect to that of reactive vitrinite in the same coal sample. Thus, the effect of pseudovitrinite on the coking properties of coals decreases with an increase in coal rank.

The positions of the lines shown in Figures 4 and 5 are not, of course, intended to represent rigidly fixed positions. However, the discontinuities were consistently documented by coking data. In fact, the recognition of these discontinuities provided the relations from which coking properties can be predicted accurately from coal petrographic analyses. Thus, it was found that the lowest and highest discontinuities, at reflectance levels of about 0.7 percent and 1.35 percent with respect to reactive vitrinite, mark significant changes in the coking properties of coals, and the specific directions of these changes were documented. On the other hand, the discontinuity at 1.0 percent reflectance turned out to be not as significant in terms of corresponding changes in coking properties.

The coking data that best support the discontinuities seen in Figures 4 and 5 include (a) the relation of reactive vitrinite reflectance to the strength of the resulting coke, (b) the relation of reflectance to the coking pressure generated during the coking cycle, and (c) the wall structures of cokes produced from reactive vitrinites.

Figure 6 shows the relation between the reflectance of reactive vitrinite and the strength of coke produced from the corresponding coals. The cokes were produced under controlled conditions in our 18 in. test oven, and the strengths were determined according to ASTM procedures. As Figure 6 shows, the metamorphic series is divided into three populations according to the stability of coke produced. The reflectance boundaries between these populations occur approximately at the 0.7 and 1.35 percent reflectance discontinuities. The lowest reflectance population of coals (0.4 to 0.7 percent) produces coke of essentially the same

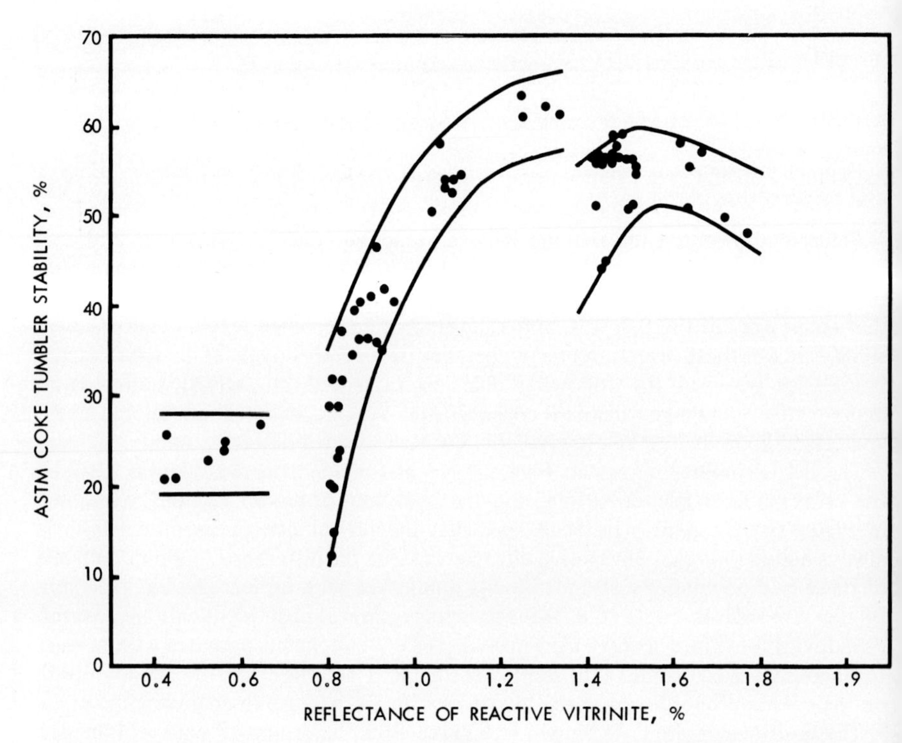


**Figure 6. Discontinuous relation between reactive vitrinite reflectance and coke strength.**